W9-BCF-755

MAN *after* MAN

Homo sapiens neanderthalensis, *once the peak of human evolution and now extinct.*

DOUGAL DIXON

MAN *after* MAN

AN ANTHROPOLOGY OF THE FUTURE

FOREWORD BY BRIAN ALDISS

Illustrations by Philip Hood

St. Martin's Press • New York

Two creatures – a single ancestor. Each is a product of 5 million years of genetic alteration and evolutionary development. Each has gone through changes – artificial and natural – imposed from outside and from within – until neither resembles in the least the common ancestral creature. The name of the ancestral creature was Homo sapiens. *It was ourselves.*

For information, address St. Martin's Press, 175 Fifth Avenue, New York, N.Y. 10010

Design by Ben Cracknell

ISBN 0–312–03560–8

First edition
First printing

Printed and bound in Italy

It is probably reasonable to conclude that, had it not been for temperature-based environmental changes in the habitats of early hominids, we would still be secure in some warm hospitable forest, as in the Miocene of old, and we would still be in the trees.

C. K. Brain

CONTENTS

2000 YEARS HENCE

5000 YEARS HENCE

10,000 YEARS HENCE

50,000 YEARS HENCE

500,000 YEARS HENCE

1 MILLION YEARS HENCE

2 MILLION YEARS HENCE

3 MILLION YEARS HENCE

5 MILLION YEARS HENCE

FOREWORD by Brian Aldiss

It has become necessary to look into the future.

There must have been a time, long past, when animals much like apes looked up into the night sky and wondered about the stars: what those pinpoints of light were, and what they were for. Only a brief while after that, the apelike things acquired language; then stories began to be told, and fantasies woven about the stars overhead. That cluster resembled a hunter and, high above, the outlines of a great bear could be discerned. Such stories, told in the Pleistocene dark, kept the bogeyman away.

Animals have no interest in stars. First speculations regarding the stars represented a revolution in thought. Speculations about the future, such as this book, mark another revolution.

Future speculation is of very recent origin. Yet today no man can call himself cultured who does not occasionally look beyond his own lifetime and his children's, if only to worry about where the cancerous growth of world population is going. Dougal Dixon's book is an ambitious attempt to view a future as far distant from us as those ramapithecine creatures whose fragmentary remains turn up in African fossil beds.

The ability to look into the future is a recently-acquired skill. It has, in fact, all been done by mirrors: there was no seeing into the future until we could see into the past. It is the ever-changing panorama of past time which we extrapolate into future time.

The business of comprehending bygone ages was a hard lesson to learn. Fossils, those coinages of past life, were always of interest to mankind. They are mentioned by Greek writers, for instance, and certainly Herodotus recognized them as being the remains of once-living creatures, understanding that their presence in the mountains of Upper Egypt was evidence that those areas had previously been under water. Lucretius, too, in his wonderful *De Rerum Natura*, pours scorn on supernatural effects and speaks of the Earth as having 'generated every living species and once brought forth from its womb the bodies of huge beasts'.

The light of reason did not always shine. Huge fossil bones later gave birth (or so we may surmise) to the legend of giants walking the Earth. The perceptions of the Greeks were forgotten. Eratosthenes, some time in the third century BC, understood well that the Earth is round, and measured its circumference with remarkable accuracy, for the latitude of Alexandria. Aristarchus of Samos, in the same period, proposed that the Earth and other planets proceeded in orbit about the sun. These perceptions were overlaid by superstition.

Greek reasoning was based on careful observation, a quality in which the Dark Ages and Middle Ages were weak. The mental world became smaller. Not until the Renaissance in the fifteenth century did learning revive. Leonardo da Vinci, for instance, studied fossils and understood their origins. He explains why leaves are found whole among rocks:

> There the mud caused by the successive inundations has covered them over, and then this mud grows into one mass together with the aforesaid paste, and becomes changed into successive layers of stone which correspond with the layers of mud.

But Leonardo did not know the age of the Earth and, in any case, accretion of knowledge is as much subject to chance and the processes of time as the fossils themselves. *Homo diluvii testis* survived as a fantasy for a while, as Piltdown Man was to do later; they were, so to say, phantom fossils.

One of the difficulties in the way of understanding the past was that for centuries the past remained obdurately and orthodoxly small. Religion got in the viewfinder. A wall rather like the walls of Jericho was built about antiquity by Archbishop Ussher, a seventeenth-century divine, who, after a careful study of the Bible, proclaimed that the world began on 26 October, 4004 BC, round about breakfast time. Precision is attractive; Ussher's calculations became dogma.

The 'walls of Jericho' begin to crumble at the beginning of the nineteenth century. What made them crumble was a tooth, retrieved from a pile of rubble in Lewes, Sussex, by a young Mrs Mantell, wife of a doctor Gideon Mantell. The Mantells took the tooth to the learned and eccentric William Buckland of Oxford, a man who ate his way through the animal kingdom and had gobbled down the heart of Richard Coeur de Lion. Buckland was a little weak on the Mantellian tooth. After some research of his own, Mantell named the erstwhile possessor of his tooth *Iguanodon*.

Buckland, meanwhile, discovered another tooth near Oxford, together with other remains, and named the fossil *Megalosaurus*.

Thus were the first two dinosaurs named. It was not until 1842 that Richard Owen defined these newly-discovered animals as a distinct group of large reptiles, and bestowed on them the label Dinosauria. A powerful new idea, a new dimension of imagination, had been born. By the time of the Great Exhibition in 1851, dinosaurs had become common property, and the notion of animals larger than elephants trundling about what became English watering places had caught the popular fancy.

Meanwhile, conceptions of the age of the Earth were being pushed out at a great rate. It spelt the fall of the house of Ussher. Evolutionary theories were current in the eighteenth century, for instance in the proposals, many of them charmingly rhymed, of Erasmus Darwin. In his *The Temple of Nature* (1803), he depicts with considerable accuracy the pageant of life from its beginnings until the arrival of mankind.

Darwin's couplets are often neat and memorable, as he intended they should be. The formation of strata of chalk is expressed in a striking image:

> Age after age expands the peopled plain,
> The tenants perish, but their cells remain.

Erasmus Darwin celebrated limestone mountains as 'mighty monuments of past delight', thus in some way looking ahead to Jim Lovelock's Gaia theory of the totality of terrestrial life as a homeostatic organism.

What Erasmus Darwin lacked was proof of his theories, the tooth found by Mrs Mantell and all the other evidences of remote and continuous life over millions of years which soon followed Owen's first christening. As geology kept pushing back the age of the rocks, it was the testimony of those rocks which supported the theory of evolution presented by Erasmus' grandson, Charles Darwin. There had to be enough time in which the whole great drama of life could be staged. Palaeontology gradually won – by a long and painstaking accumulation of facts by numerous people, learned and not so learned.

We now know that life on the planet is no less than 2500 million years old, whereas the age of the Earth is accepted as being something more than 4500 million years.

It was my good fortune as a boy of seven to be given an imposing volume entitled *The Treasury of Knowledge*. There for the first time I learned of evolution and of the ages preceding ours. So enamoured was I of the story of the creation of the solar system, of the dawn of life, of the dinosaurs, and of those early men – like us, unlike us – that I gave lessons on the subject when at preparatory school, at one penny a time. Although I do not recollect ever being paid, I recall the pleasure we all had drawing brontosauruses and shaggy Neanderthal men.

That precious book is still in my possession. It was published in about 1933 (no actual date printed). Nowhere does it give the ages of the various epochs of past history. A question mark still hung over that subject in the years before carbon-dating and an understanding of the nuclear nature of the sun. In one lifetime we have progressed from that grey area to knowing (or believing we know) how the universe itself came into being – though some doubt remains about the first few seconds of that event.

Until we could look into the past, until the past was seen as a story of continuous development or change, with the mutability of species which that implied, the future remained blank. It gave no credible reflection. This we can see if we read romances of the future penned before evolutionary theory became a reality in human minds. Futures were like the present but more so.

Mary Shelley's *The Last Man* of 1826, for instance, is set at the end of the twenty-first century. It is a bold stroke, and some play is made with travel by air balloon and revolution in England; but the Turks are still causing trouble at the eastern end of Europe. When a plague commences to wipe out all of humanity, no attempt is made to introduce innoculation or vaccination, although that would have been a reasonable proposition in the 1820s. The novel is full of interesting reflections; but the motive power which evolution could supply is absent.

It was not until 1895 that readers could take up the first novel to be formed by evolutionary thought, as a waffle is shaped by the pattern of the waffle iron. *The Time Machine* was written by a pupil of Thomas Huxley, Darwin's great protagonist, H.G. Wells. In this marvellous narration, Wells sketches out aeons of future time. It is part of his design that – unlike the epochs in *The Treasury of Knowledge* – everything has a date. The date at which the time traveller eventually arrives is 802,701: not, in fact, a credible date for the end of the Earth by today's standards, but one well designed to seem reasonable to the book's first readers, who had enough other marvels to cope with. Indeed, it is difficult to realize now just how subversive the book must have seemed to many at that date, for a gloomy picture indeed is painted of the bifurcation of society into Morlock and Eloi to which Victorian society is depicted as heading. Evolution is shown as not working on behalf of mankind, as was then popularly imagined.

And, of course, our species is shown as mutable, as transitory.

As the time traveller travels through time into a distant

future, he observes that 'The whole surface of the earth seemed changed – melting and flowing under my eyes'. This is a man who has read Sir Charles Lyell's *Principles of Geology*. 'I saw great and splendid architecture rising about me, more massive than any buildings of our own time, and yet, it seemed, built of glimmer and mist.' It is not only man's achievements, but mankind itself, which proves transitory, a thing of glimmer and mist.

Without a fresh understanding of the past, without its decipherment, *The Time Machine* could not have been written; or, if written, could not have been deciphered.

Following on from Wells, we have had many visions of the future. Whether mechanical, trivial, or profound, they all rest on the findings of the nineteenth century; all work as reflections of our understanding of the preceding millions of years.

As much is true of Dougal Dixon's book. Yet it impresses me as being startlingly original, perhaps the progenitor of a new breed, future-faction. It eschews the trappings of fiction upon which Wells seized. It presents itself as a straight record of the future, the future over the next 5 million years. It is Darwin, Lyell and Wells rolled into one. They would like this book, and be horrified by it: for we have, after all, travelled a long way since their day, and supped on horrors beyond their resources. We have lived through an age (well, men felt much the same in 1000 AD, though for different reasons) when we have almost daily expected the world to be terminated.

So here is the mutability, with human flesh a thing of glimmer and mist. *Man After Man* is a drama of the oncotic pressure of time on tissue. Dixon does not tell us of the things his caravanserai of creatures believes and thinks; it is enough that we know what they eat. For one of the revelations brought home by evolutionary theory is that we are a part of the food chain, along with pigs, broiler fowls and the tasty locust.

Of course the prospect is melancholy as well as fascinating. This is one of the characteristics of futurology. After all, we are looking at a period long after our own insignificant individual deaths. Everything we are asked to consider here reinforces the fact that our world and all we cherish in it is gone. We are one with Tutankhamun and Archbishop Ussher. Other beings possess the field.

Consider Knut who, Dixon tells us, lives a mere 500 years from now. Knut's seems a lonely life. He lives in a wilderness of tundra. He subsists on a diet of mosses, lichens, heathers, and coarse grasses. He has been adapted, so he finds his diet palatable and nourishing. But the question arises in our minds: do we not find a little frightening and alien this inheritor of our world – and where did all the toast and marmalade go?

We ourselves like – need – a coarse mental diet. We pass for human, but perhaps only among ourselves. Part of us is sane but, at times of crisis, and not only then, an instinctive drive takes over. We seek to set aside the human aspect by use of drink, drugs and other means of escape, as if being human was as yet too much for us. We have a hearty appetite for apocalypse, as the history of the twentieth century shows.

With this appetite goes an obsession with the future. The futures we visualize are generally dystopian. Dixon's is science-based, but proves distinctly ahuman. Sombre, I would call it. And sombre was also a word that occurred to Thomas Hardy when he considered the change in taste of our modern age. Hardy was a pall-bearer at Darwin's funeral, and his writings are steeped in evolutionary thought, from *A Pair of Blue Eyes* to *The Dynasts*, the great supernatural drama he wrote in the early years of this century. In *The Return of the Native*, he reflects on such matters:

> Men have oftener suffered from the mockery of a place too smiling for their reason than from the oppression of surroundings oversadly tinged. Haggard Egdon appealed to a subtler and scarcer instinct, to a more recently learnt emotion, than that which responds to the sort of beauty called charming and fair.
>
> Indeed, it is a question if the exclusive reign of this orthodox beauty is not approaching its last quarter. . . . Human souls may find themselves in harmony closer and closer with external things wearing a sombreness distasteful to our race when it was young. The time seems near, if it has not actually arrived, when the chastened sublimity of a moor, a sea, or a mountain will be all of nature that is absolutely in keeping with the moods of the more thinking among mankind.

Hardy there shows his prophetic sense. We might go on to say that chronicles of change which impress on us the transitory nature of our lives and our civilization are also in keeping with the mood of the present. The current obsession with the future may also pass away in time; but for now – just for now – Dougal Dixon has the right idea.

Brian Aldiss

INTRODUCTION – EVOLUTION AND MAN

Evolution is the process that brought us to where we are today.

It started about 3500 million years ago, when the first living thing, probably a single complex organic molecule in the form of a long chain, began to reproduce itself. It did this by latching onto simpler molecules dissolved in the water around it, until it built up a mirror image of itself. The two parts then split apart to become two identical complex molecules. Each of these had the same power of attracting simpler molecules and building up a mirror image – similar to the way in which viruses reproduce themselves.

The building up and the splitting took place untold millions of times. Inevitably on occasion the mirror image so produced was not accurate. As a result the new molecule had slightly different properties from the old, and may not have been so efficient at reproducing itself. In this case the changed molecule – the mutation – stopped reproducing and died out.

However, the occasional mutation arose that actually helped the molecule to reproduce itself. The mirror images - the offspring - of this mutation then survived. This is the basis of the process that we call evolution.

After millions of beneficial chance mutations the single molecule became more and more complex, if complexity ensured a more efficient reproductive process. The molecule changed from a virus-like entity to a living cell, in which the reproductive molecule or molecules were encased and protected by an outer membrane. This resembled one of our modern bacteria.

The chemical reactions that enabled early molecules to reproduce themselves may have been insufficient to power the reproduction of more advanced creatures, and other energy sources developed that allowed the absorption of energy from sunlight and the use of this energy to build up raw materials for reproduction. The first single-celled plants had evolved.

Other mutated cells did not use the sun's energy. Instead they digested the cells that did, and so used the energy already stored. These were the first animals.

Eventually creatures evolved that consisted of more than just a single cell. This came about either by cells reproducing themselves and then failing to split, or by several cells coming together. Whichever it was, if the multi-celled creature were more efficient, then it survived and reproduced in its multi-cellular form.

With the increasing complexity, the different cells in a single creature evolved to have different functions. Some cells were involved in sense, helping the creature to find food or light. Other cells were involved in locomotion, in moving the whole creature towards its food or its light source. Others were involved in digestion, others in reproduction, and so on.

The different masses of cells are what we call tissues, and the structures that they form, each with a different function, are called organs. An entire creature (made up from molecules that make cells, that make tissues, that make organs) is called an organism.

At an early stage the pathways of evolution began to branch, and different types of organism developed. Wherever there was a food source that could be exploited, evolution produced an organism able to exploit it. Such a process is called adaptive radiation, and we can see it at work today.

Many species of finch live in the Galapagos Islands, off the west coast of South America. These all evolved from one type of seed-eating ground finch that came over from the mainland, and spread to all the islands, each with different habitats and food sources. The finches on each island evolved to take advantage of their particular habitat. As a result there are now many species of finch on the islands, including heavy-beaked forms that eat seeds, short-beaked forms that eat buds and fruit and long-beaked forms that eat insects.

Environments are not stable; they change for one reason or another. When this happens, a creature evolved to live in a particular way in a certain environment becomes extinct. For example, if all the insects on the Galapagos Islands died out, then the long-beaked finches would become extinct: a process known as natural selection. If the insects became extinct, their places would be taken by another creature, and some other bird would evolve to eat that.

Evolution produces specific shapes of animals to live in particular environments. Grass is tough to eat, so an animal that eats grass needs strong teeth and a specialized digestive system. Grasslands are wide open areas in which danger can be seen coming from a long way away, and there are no hiding places. A grass-eating animal, therefore, tends to have long running legs, as well as strong teeth, and a long face so that its eyes are above the level of the grass while its head is down eating. This gives us the shape of the antelope – the typical grass-eating animal of Africa.

However, the grasslands of Australia have evolved a quite unrelated grass-eating animal – the kangaroo. There seems little resemblance between this and the antelope of

Africa. It does, however, have the same long face with similar grass-grinding teeth; and the legs are long and built for speed, albeit in a bounding rather than a running gait. This development of similar features in unrelated animals in response to similar environmental conditions is what is known as convergent evolution. It accounts for the similarities between seals and sealions, aardvarks and anteaters, ants and termites, vultures and condors.

A similar phenomenon is parallel evolution. In this, two branches of the same family tree develop along similar lines independently of one another. For example, the kit fox of North America and the fennec fox of Africa are both small, with a sandy pelt and large ears. The ears act as cooling vanes and prevent each animal overheating in its desert environment, and the pelt is camouflage. Both are descended from a more conventional fox-like animal, but each has evolved separately to live in different deserts.

The different colours and patterns in animals can also be attributed to evolutionary processes. Animal patterns may camouflage them: on the other hand they may, like the skunk, have striking colours that warn a would-be attacker that the owner is poisonous. Some animals mimic others, as when a harmless king snake develops the spectacular pattern of the poisonous coral snake, and consequently turns away potential enemies. All these have developed because the animals concerned have benefited from them, have survived and have gone on to reproduce.

Throughout the world and throughout time, animals and plants have changed in response to the changes in the environment.

One species has broken with this tradition. Within the last million years or so the human species *Homo sapiens* evolved. It has come all the way from molecules to its present form in 3500 million years by the workings of evolution. Now, within the last few millennia, intelligence has developed, and with it cultures and civilizations. The species has spread not by changing to adapt to the environments it found but by changing the environments to suit itself. Instead of developing furry pelts and layers of insulating fat to adapt to cold conditions, it manufactures artificial coverings and uses available energy supplies to generate heat for the body. Instead of evolving heat radiating structures such as big ears to adapt to hot conditions, it manufactures refrigeration and air-conditioning systems, again using available energy supplies. Instead of developing speed and killing strategies that allow it to hunt a particular food, it builds machines to do it. By using its intelligence it can exploit all food supplies in all environments without having to change itself.

Medical science eliminates much of the effects of natural selection: no longer does an individual not particularly well adapted to the environment die out before being able to reproduce.

Under natural conditions not all offspring of a species survive, and this is reflected in the birth-rate. Thanks to medical science, more offspring of *Homo sapiens* survive than ever could before, but this has not been reflected in a corresponding drop in the birth-rate. As a result the populations of *Homo sapiens* are growing without the refining and modifying processes of natural selection.

Evolution as we know it for *Homo sapiens* has stopped. However, this does not mean that the process of change has necessarily stopped.

As science develops, the reproductive molecules – the genes – that exist within every cell of the human body are becoming better and better understood. When *Homo sapiens* finally appreciates which parts control the development of which features, then the possibility exists for modifying the process. A stage will be reached when one gene can be suppressed, another encouraged, with yet another created from new. A human being with particular features, following a particular preconceived plan, may be born from modified sperm cells and ova. Without the natural processes of modification, this unnatural process is the only way of developing the species into new forms to face the problems that await it in the future: problems generated by overpopulation, over-use of natural resources and pollution.

Genetic engineering

The mechanics of genetic engineering are already complex, yet in their current state they are primitive compared to what will undoubtedly be possible within a few decades.

The reproductive molecules that lie at the nucleus of each cell of a living organism are in the form of long structures called chromosomes. These chromosomes are made up of the chemical substance DNA. Its shape is best imagined as a long ladder that has been twisted along its length. Each rung of this ladder consists of two compounds, called bases, locked together. There are only four different kinds of bases: thymine, cytosine, adenine and guanine, referred to as **T, C, AS** and **G**. A **T** always unites with an **A,** and a **C** always with a **G**. The sequence of these base pairs along the twisted ladder of the chromosome is almost infinitely variable – there are something like 6,000,000,000 bases in a full set of human chromosomes.

A chromosome is often described as a page in an instruction manual. Each base pair, or rung in the ladder, represents a letter of the alphabet, and the arrangement along the ladder gives 'words' and 'sentences'. Each understandable instruction so formed gives a gene. The genes in a single cell produce the total information needed for the growth of the entire organism.

When an organism grows and develops, it does so by multiplication of cells. Each cell splits into two complete cells. When this happens, each chromosome in the cell actually splits down the middle. The uprights of the twisted ladder pull away from one another as the rungs split into two along the joins between the bases. What happens then is that these two half-ladders build up two complete ladders by attracting free bases made up from the chemicals drifting in the cell. As a result, when the cell splits into two each new cell carries exactly the same set of gene instructions.

The exception to this process is in sexual reproduction. Reproductive cells carry half the normal number of chromosomes. Two half-cells unite during fertilization to produce one cell with the full number. This new cell is a unique mix of genes, half from the mother and half from the father. This cell then divides in the usual manner until the entire organism is built up, following the instructions now carried in every cell.

The big mystery now is this: how do the genes – the pattern of base pairs along a chromosome - actually work? How do they control the construction of an organism?

The idea behind genetic engineering is to manipulate natural processes. In some way genetic instructions along the chromosomes in a cell have to be identified then changed so that as the organism grows, it is to a new set of instructions. Since all the materials involved (cells, chromosomes, molecules) are microscopic, a whole new technology has to be applied.

Viruses can do it. Viruses essentially consist of a mass of their own DNA encased in an envelope. When they infect a cell they attach themselves to the cell's wall and inject their DNA through it. In the cell's interior the invading DNA breaks down the cell's chromosomes and rebuilds the material into copies of itself.

For genetic engineers to do the same, they would first of all have to break in through the cell wall, then break down the DNA of the nucleus and reassemble it in the desired way. Alternatively, they could cut out segments of the DNA strand, segments that correspond to particular genes, and replace them with DNA segments already prepared. This would be done by chemicals that have specific biochemical reactions – enzymes – some of which have been found to have the ability to cut DNA strands.

The greatest experimental successes so far have been with bacteria. These single-celled creatures have cell walls that can be softened by chemical solutions so that new DNA can be placed inside. The double helix of the original chromosome can be chopped up using enzymes, and new DNA can be inserted. The broken ends of the DNA strands have one side longer than the other, exposing a sequence of bases. If the introduced DNA segment has matching bases exposed at its end the two DNA pieces will unite, **T** to **A,** and **C** to **G,** and produce a complete chromosome. This technique is known as gene-splicing.

Before any of this can be attempted, however, the whole gene pattern has to be mapped. At the moment only about 100 human genes have been identified and interpreted; but, since genetics has only been in existence for a century, and the structure of the chromosome has only been known for about four decades, and scientific advance in this area is increasing exponentially, what was speculation about genetic engineering is quickly becoming fact.

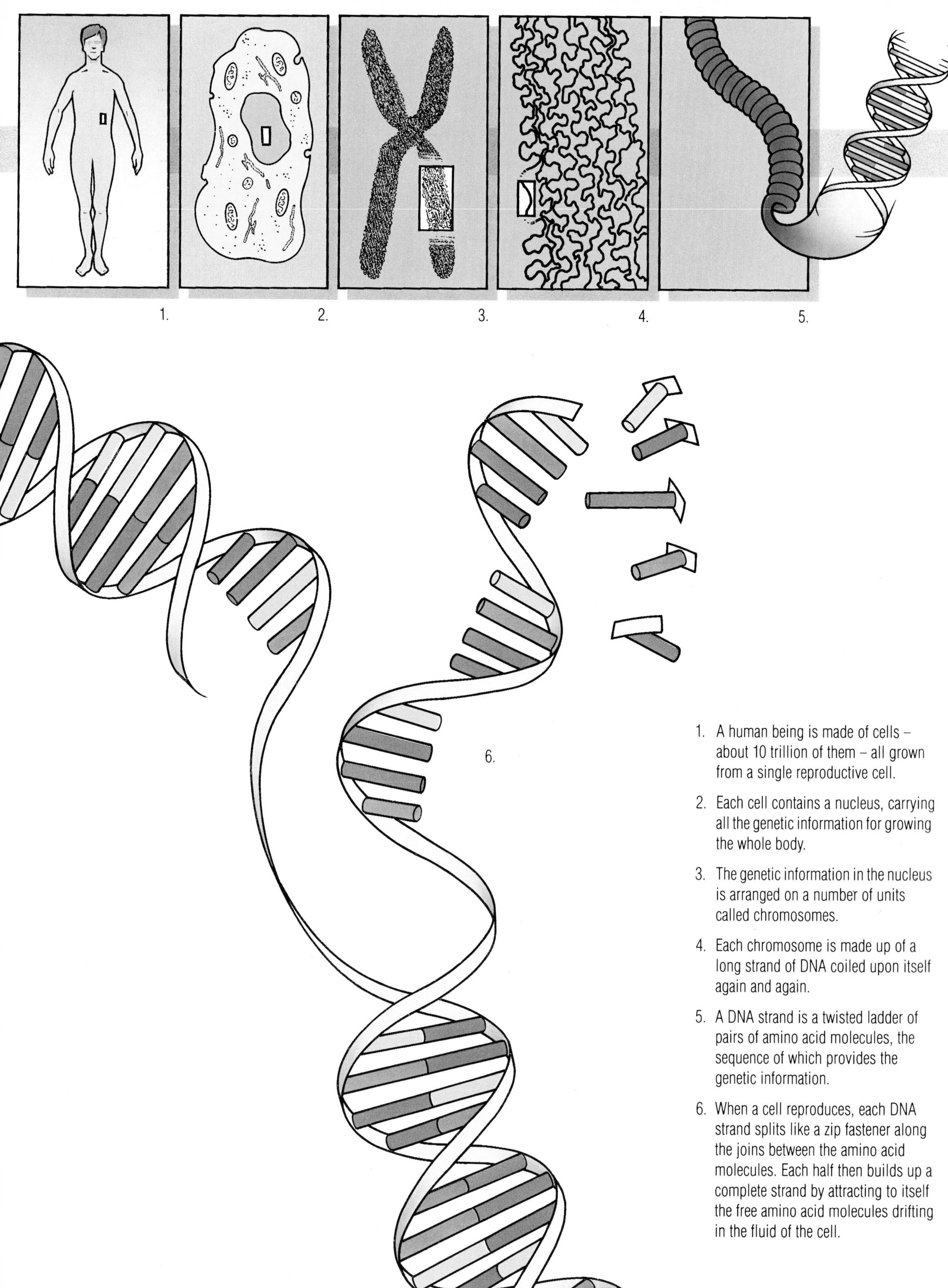

1. A human being is made of cells – about 10 trillion of them – all grown from a single reproductive cell.
2. Each cell contains a nucleus, carrying all the genetic information for growing the whole body.
3. The genetic information in the nucleus is arranged on a number of units called chromosomes.
4. Each chromosome is made up of a long strand of DNA coiled upon itself again and again.
5. A DNA strand is a twisted ladder of pairs of amino acid molecules, the sequence of which provides the genetic information.
6. When a cell reproduces, each DNA strand splits like a zip fastener along the joins between the amino acid molecules. Each half then builds up a complete strand by attracting to itself the free amino acid molecules drifting in the fluid of the cell.

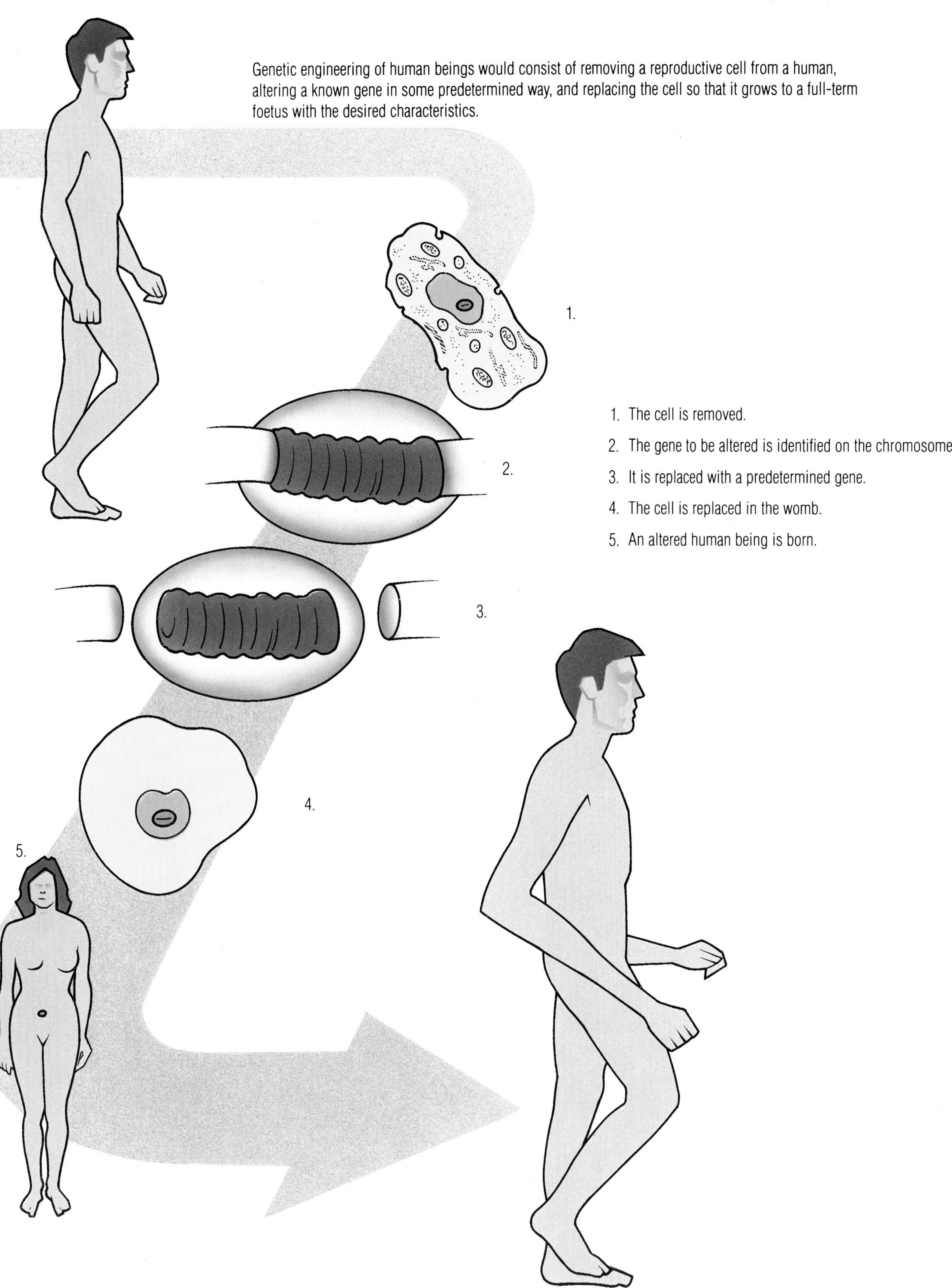

Genetic engineering of human beings would consist of removing a reproductive cell from a human, altering a known gene in some predetermined way, and replacing the cell so that it grows to a full-term foetus with the desired characteristics.
1.
2.
3.
4.
5.
1. The cell is removed.
2. The gene to be altered is identified on the chromosome.
3. It is replaced with a predetermined gene.
4. The cell is replaced in the womb.
5. An altered human being is born.

PART I:

IN THE BEGINNING – The Human Story So Far

8 MILLION YEARS AGO

Her ancestors lived in the treetops that once covered the area. Indeed her relatives still live in the forests of the steamy lowlands, climbing the branches, eating the soft fruits and grubs; her way of life is, however, completely different. Hers is a dry landscape of yellow grass, with brown and black thickets of hardy thorn trees.

Her woodland diet is different, too, because there are no soft fruits and juicy buds or grubs here. Solid nuts and tough seeds are her main foodstuffs, and when there is nothing else she makes do with coarse roots and tubers. Hard-shelled insects and dry lizards abound, and she often extracts what little nutrition there is from these. Her jaws and teeth reflect the fact that she has to eat more than her ancestors did to gain the same amount of goodness, and she has to chew it more thoroughly. Accordingly, her front teeth have become smaller to make room for broad and flat back teeth that grind down masses of coarse food. This has not happened suddenly, but has developed over thousands and thousands of years. Those who study her remains will give her a name. They will call her *Ramapithecus*.

The other animals that live here show the same specializations in their teeth. Pigs and antelope feed on low-lying plants, and giraffes browse the higher trees. These too have broad back teeth; but she has a long way to go before she is as well-adapted as they are.

For one thing, the grasses are very tall, and when she is on the ground she is lost and cannot peer over them. There are fierce hunting beasts around, too, so she needs to climb the trees for safety as well as to see distances. The other animals run away when threatened, but she does not have the speed, running on all fours on short limbs.

Stiffly she pushes herself to her hind feet, and sways unsteadily for a time. Now she can see over the top of the grass, and, what's more, she feels cooler. Less of her back is exposed to the hot sun, and the cool breeze that she now feels soothes her neck and chest (overheating was not a problem in forest shade). The more comfortable temperature, however, is counteracted by discomfort in her legs, as this is not a natural pose for her. Maybe she can move more quickly like this, with only two feet touching the ground. She tries but her legs are not strong enough, and are the wrong shape for this to work. Her body naturally topples forwards, and she cannot move her hind legs quickly enough to stay upright.

She descends once more onto all fours. No. She will have to stay near the trees if she wants to survive.

3 MILLION YEARS AGO

The climate is much drier now, and the scenery has changed considerably. The continent has been moving, gradually splitting the landscape across with faults, while elongated slabs have slowly subsided forming long, deep, rift valleys with strings of shallow lakes in their floors. Molten material has been brought up from the Earth's interior, and active volcanoes line the edges of the rift. Grasslands have spread everywhere and there are many clumps of trees, but no continuous forest.

At the edge of one such clump a small creature drops from a tree to the ground; and then stands upright. He looks around for danger and, seeing none, grunts a signal. The dozen others who drop from the branches and cluster around him include other males, much smaller females (some with babies) and children – it is a large family group.

Food has become sparse in their thicket, and they are moving. Further down the valley a patch of green by a lake holds out some hope. With a confident stride, they march downhill, leaving footprints in the volcanic ash that carpets the whole area from the last eruption. Their stride and their stance show that their legs have developed considerably in the last 5 million years. From permanently-bowed structures only good for climbing trees, their legs have developed into straight limbs that can carry their bodies vertically. Their arms, however, have changed little during that time: they still have the curved fingers for grasping branches, and the shoulder-socket angled upwards allowing a high reach, both features of a tree-living way of life.

If the landscape becomes much drier, though, and the trees more sparse, beings that are better adapted for a ground-dwelling existence will be more likely to survive than this partially tree-living creature, *Australopithecus afarensis*.

That time is not far off.

2.5 MILLION YEARS AGO

Volcanoes still bubble; grassy plains still spread along the rift valleys, but now only isolated umbrella-shaped trees and low thorn thickets break up the yellow of the landscape. Down by the edge of the lake a pack of large hyenas has brought down an animal that looks like a short-necked giraffe with moose-like horns, and are tearing its corpse apart.

In one mass of bushes a number of heavy-looking beasts forage amongst the thorny vegetation for leaves and ber-

Ramapithecus – *ancestor of apes and humans.*

Australopithecus robustus – *the vegetarian dead-end.*

ries. If it were not for their upright stance they would be mistaken for chimpanzees, as they have the same heavy bodies and the same deep jaws with massive teeth. These also belong to a species of *Australopithecus*, called *Australopithecus robustus*, and they are perfectly at home here as they contentedly chew any piece of vegetable material they find.

Suddenly the nearby grass erupts. About a dozen screeching figures run at the feeders. They look much like the others, but are more lightly built and their faces do not have such a heavy-jawed look. They belong to another species, *A. africanus*.

The feeders stop eating and snarl back, staring defensively at the newcomers and showing their teeth and gums. They are not to be chased away from their feeding ground. The attackers halt in their assault; their intended victims seem more determined than they anticipated.

The attackers back away slowly, keeping up their aggressive noises and trying not to appear vulnerable, then regroup some distance away. The berries of the thicket are lost to them.

They turn their attention to the hyenas feeding down by the lake and, as a group, charge them. The hyenas are startled by this sudden assault and in a panic they abandon their kill. The attackers gather around the corpse, some of them tearing at the meat while others stand guard, waving sticks and snarling at the cheated hyenas.

These creatures can eat meat as well as plants, and can combine forces in order to procure it. Their larger relatives in the thicket continue munching their berries – meat-eating and co-operative hunting is not for them.

1.5 MILLION YEARS AGO

It seems the same place, for the landscape has changed very little; though the climate is now much cooler. Large chimpanzee-like creatures still forage for berries amongst the bushes. These creatures, however, are larger than the earlier berry-eaters, and have very heavy jawbones. Later, anthropologists gave them various names such as *Zinjanthropus*, Nutcracker Man, before deciding that they were members of the earlier *A. robustus*.

Not far away several very much smaller ape-like beasts, evolved from the earlier *A. africanus*, carry a dead antelope between them. That is not all they carry: they have stones that have been chipped into edges, points and blades, for these creatures are tool-makers, and as such they have a culture, later referred to as the Palaeolithic, or old stone age. Their scientific name reflects this tool-making skill: it is *Homo habilis*, meaning 'handy man'.

The two groups pass very close to one another, but totally ignore each other's presence. Now they have evolved in such diverse directions, they no longer compete for the same food.

500,000 YEARS AGO

She is a member of the first group of humanoid creatures to move out of Africa and spread across Europe and Asia. She crouches in a cave entrance in what will be known as China; but far away, in places that will be called Spain, Java and Tanzania, there are beings just like her.

If she stood up she would be seen to be very similar to a twentieth-century human, but with a heavier jaw, protruding eyebrows and a flat forehead. Her upright stance gives her species the name *Homo erectus*.

As she watches the hunters drag home the slain bison, while other females carry back their handfuls of hackberries and pine kernels, her thoughts are only on the food that they bring, and how this food is to be prepared.

Co-operation with others and skills learned from her parents provide her with food. With a stick, she stirs the powdery whiteness in the fire pit before her, uncovering the deep red glow. She adds dry twigs to bring the glowing embers to life. She cannot remember when or how the fire started but hers is the responsibility for keeping it going. It is a heavy responsibility, too, since fire makes the meat tender enough to eat easily, its smoke preserves what meat they do not eat immediately, and its frightening light keeps away the fierce night animals.

She knows that it is her responsibility because the group of 23 who occupy the cave have 'talked' it over – not in words but in significant sounds that mean something to those in the group – a long stride on the road to civilization.

15,000 YEARS AGO

A horse develops before him. Red soil from one part of his stone dish has been applied with a pad of moss to the cave

Australopithecus africanus – *the adaptable survivor*.

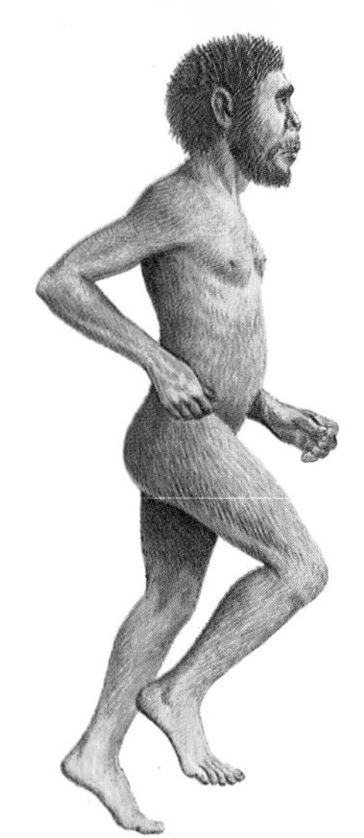

Homo habilis – *the tool-maker*.

wall to block in the basic shape. Now he takes soot and smears it along the figure's back, pointing up its ears. The same black pigment goes into making the legs and the hooves.

In the confined space, and by the flickering light of his flame, it is difficult for him to stand back and appreciate his work. He knows, however, that he has done it to the best of his ability, and this gives him a deep satisfaction.

Squeezing through the narrow limestone passage towards the cave mouth he passes other paintings. Bulls, reindeer, bison and rhinoceros have been depicted there since long before his time.

He blows out his flame and stands, dazzled, on a limestone shelf looking down the hill at the wooded gorge below. Smoke rising against a far cliff shows where his people live, sheltered against the coming winter blast beneath the overhang.

He belongs to the species *Homo sapiens*, subspecies *sapiens*, and there are probably no more than 10,000 like him in the area that will one day be known as central France. Further to the north, on the tundra plains of Germany, his cousins *Homo sapiens neanderthalensis* are now extinct, either wiped out in the latest surge of the ice age, or else so interbred with the more successful *Homo sapiens sapiens* that their characteristics have disappeared in their offspring. It is *Homo sapiens sapiens*, or Cro-Magnon man, with his artistry and his advanced Palaeolithic culture, who will be the ancestor of mankind to come.

5000 YEARS AGO

The river valley has always produced the best plants and, since most food comes from one plant or another, the river valleys of northern Europe are well settled. With the knowledge that plants grow from seed, the people of the settlement have gathered seed and planted it in the fertile valley soil. When the plants are ripe they are cut down with stone-bladed sickles, and the seeds ground down to flour by rolling them between coarse stones.

What can be done for plants can also be done for animals. On the cold plains to the north people still follow migrating herds of reindeer, so that meat is always available; but the settlers can do better than this. Their animals – their cattle, sheep, goats and pigs – are kept penned near the settlement so that meat, wool and milk are constantly accessible.

As a result, for the first time in history substantial houses can be built, on frames of tree trunks, hewn by the stone implements, walled by dried clay and sticks. Straw, left over from the grain harvest, goes into making the roof. Now there is also time and opportunity for pottery and horn ornaments to be crafted.

It is the era known as Neolithic, or new stone age. The cultivation of plants and the domestication of animals have both heralded this new culture. It will not be long now before the settlers, with their more stable lifestyle and the time to apply their minds to abstract problems, learn to smelt and use metals – first bronze and then iron – and this knowledge will spread throughout most of the populated world.

2000 YEARS AGO

Lucius Septimus chews his twice-cooked bread at the entrance of his hide tent, having cleaned his iron weaponry and his armour. Out there, in the rain, the grey choppy sea that beats against the northern limit of Gaul is an uninviting sight. The wild Britons of the lands to the north have been a thorough nuisance, giving constant aid to rebellious Gauls and holding up the establishment of Roman civilization in these northern lands.

Also, it is said that there is great mineral wealth to be had there. Stories abound of wealthy metal merchants making their fortunes by plying these dangerous waters.

Certainly the military victory achieved there by the late Julius Caesar was small; but the talk is that other invasions are planned. He certainly hopes not. He would rather be serving in newly-annexed Aegyptus at the other end of the empire.

Only the generals and the officers in the big tent at the end of the row know what the long-term plans of the new emperor Augustus are. Lucius merely goes where he is told, and fights where he is told. He feels lucky to be a part of the great nation of Rome: a nation that controls practically the whole world and will do so for ever.

1000 YEARS AGO

Empire after empire developed around the Mediterranean sea and spread across Europe, Africa and Asia, clashing

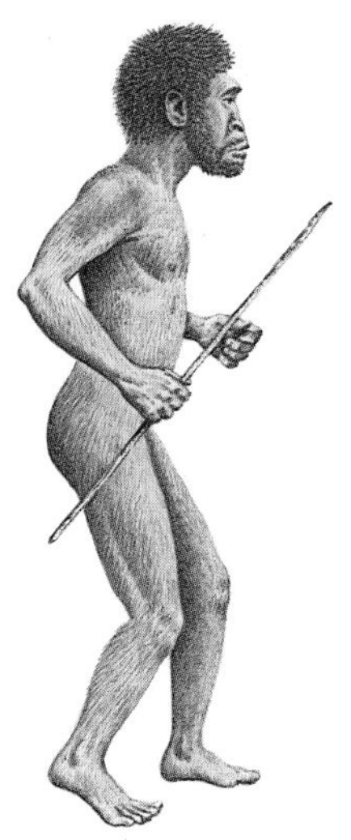

Homo erectus – *the fire-maker.*

Homo sapiens neanderthalensis – *our less successful cousin.*

with the other empires found there. Then they collapsed; and usually the culture and technology generated with each empire collapsed with it.

Eyjolf Asvaldsson understands little of this. He is about to sail home, guided by the stone that seeks the north star. He does realize, however, that places visited by long ships during the summer raids seem to have different histories, and display different ruins.

Almost everywhere in the world shaven men teach the Christian faith and vehemently denounce the sacred names of Thor and Odin; and everywhere the people are adopting this faith – even some of Eyjolf's own people. In this country, the Arab Kingdom of Spain, is a mixture of religions. Dark-skinned peoples who scorn the Christian religion have been settling here for a long time, alongside Christian people. They worship God in domed buildings, surrounded by spindly towers. What's more, they are gardeners and poets, and have a technical knowledge that is lacking elsewhere.

Eyjolf's abiding memory of the last raid is of a tower with sails. Ships, like his own, use the wind; they catch it in their sails and it drives them along. These people, however, use the wind to turn wheels and grind grain.

500 YEARS AGO

It is 69 days since they set out from Palos, and all that time they have been sailing westwards, except for a brief stop for provisioning in the Canary Islands. Now they have arrived, in India.

Pablo Diego chides himself for mistrusting the captain. There was no way of telling whether or not the voyage was foolhardy. They just kept sailing westwards – totally the wrong direction for India – to the edge of the world, possibly to be enmired by sticky seaweed or eaten by sea monsters. They could tell how far north or south they were, by measuring the angles of the stars, but there was no way of telling how far west they had sailed. Several times he and the crew were on the verge of mutiny.

They were wrong, however, and now here they are, safe beneath the palm trees on the warm beach, while offshore the three proud ships lie resting at anchor. It is the Indians that puzzle Pablo. Evidently this is *not* the mainland of Asia, but one of the outlying islands, possibly the Japans.

But where are the fabulous treasures, the gold and jewels that have been promised? Friendly or not, the gifts that the Indians bring are rubbish – beads and strangely-coloured birds. Nevertheless, they do have gold rings in their noses; so there is wealth somewhere.

If there is, why are the Indians not using it? They seem to have nothing, living in grass huts and growing strange plants for food. That does not worry Pablo. The captain has said that after a brief rest they will sail around more of these islands. He can be sure that further to the west is the main continent – a civilized continent of civilized people who know what to do with their wealth.

100 YEARS AGO

The train rattles out from between the narrow paper houses, sending up thick clouds of black smoke that settles as soot on the ornate carvings of the eaves, then coughs its way along the low embankment between the flooded fields of rice towards the distant cotton mills. If there is anything that emphasizes the changes that have come to Renzo Nariaki's beloved Nippon it is this. He is an old man now and he can still remember his place in the feudal society of the Tokugawa Shogunate before it was overthrown.

Then, with the civil war and the emplacement of the emperor Meiji, the barbarians who had long been attempting to gain a foothold finally flooded in. They arrived at the request of the new emperor, and changed everything.

They were altering all aspects of society. At least he still had an emperor, but the government was now like that of a place called France. They still had a navy, but run along the lines of the British navy. Their industry was being reorganized into the American style; while the army was no longer the army of the Samurai – it was now like the army of Germany.

The train has disappeared into the dark mills now, ready to pick up a heavy load. The traditional road transport could never have handled the volume of goods now being produced. It is probably like this all over the world, thinks Nariaki. The foreigners are imposing their way of life everywhere.

Or perhaps we are absorbing the foreigners' way of life? Time will tell.

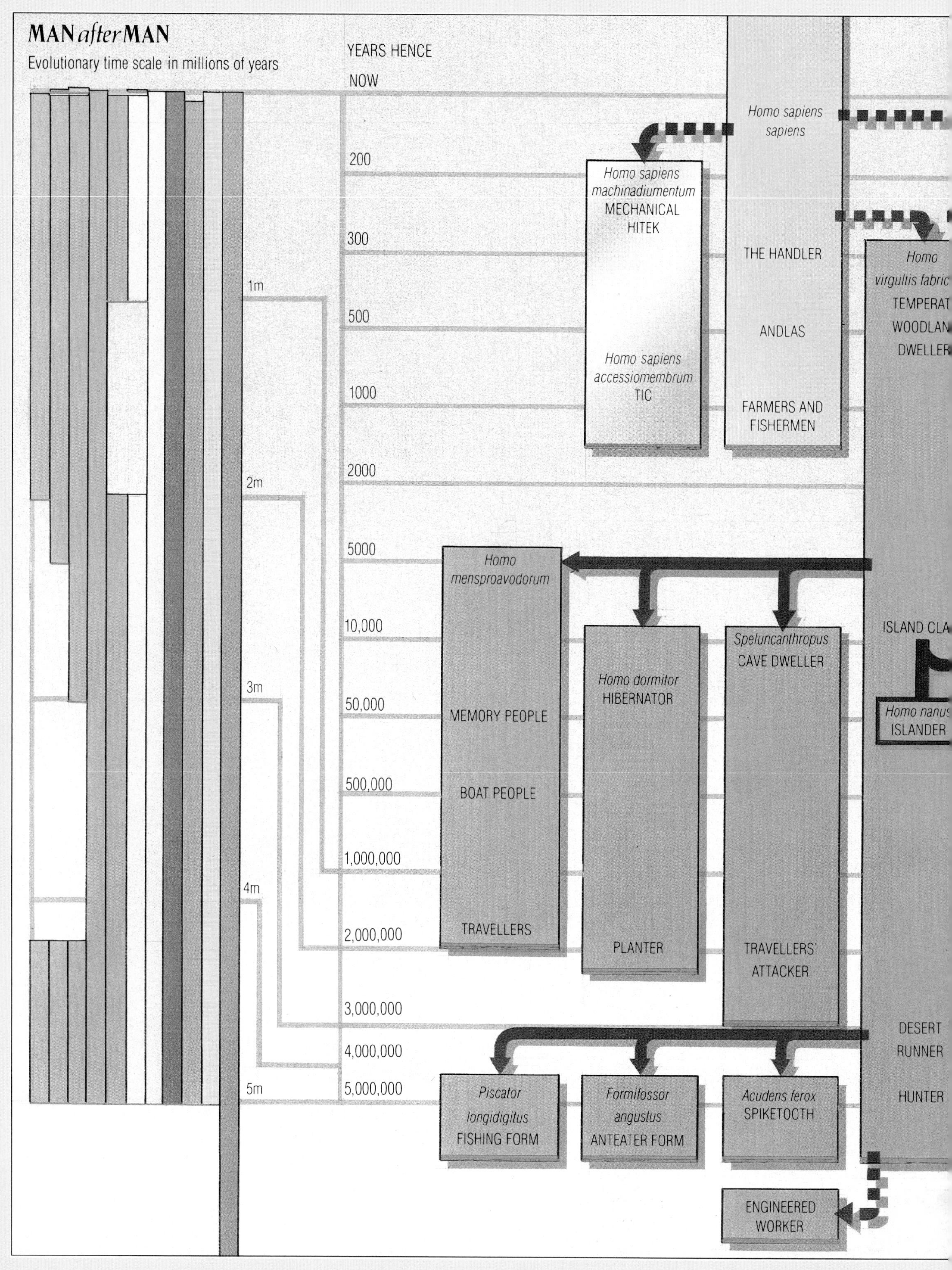

MAN after MAN
Evolutionary time scale in millions of years
YEARS HENCE
NOW
200
300
500
1000
2000
5000
10,000
50,000
500,000
1,000,000
2,000,000
3,000,000
4,000,000
5,000,000
1m
2m
3m
4m
5m
Homo sapiens sapiens
Homo sapiens machinadiumentum
MECHANICAL HITEK
Homo sapiens accessiomembrum
TIC
THE HANDLER
ANDLAS
FARMERS AND FISHERMEN
Homo virgultis fabric
TEMPERAT
WOODLAN
DWELLER
Homo mensproavodorum
MEMORY PEOPLE
BOAT PEOPLE
TRAVELLERS
Homo dormitor
HIBERNATOR
PLANTER
Speluncanthropus
CAVE DWELLER
TRAVELLERS' ATTACKER
ISLAND CLA
Homo nanus
ISLANDER
DESERT RUNNER
HUNTER
Piscator longidigitus
FISHING FORM
Formifossor angustus
ANTEATER FORM
Acudens ferox
SPIKETOOTH
ENGINEERED WORKER

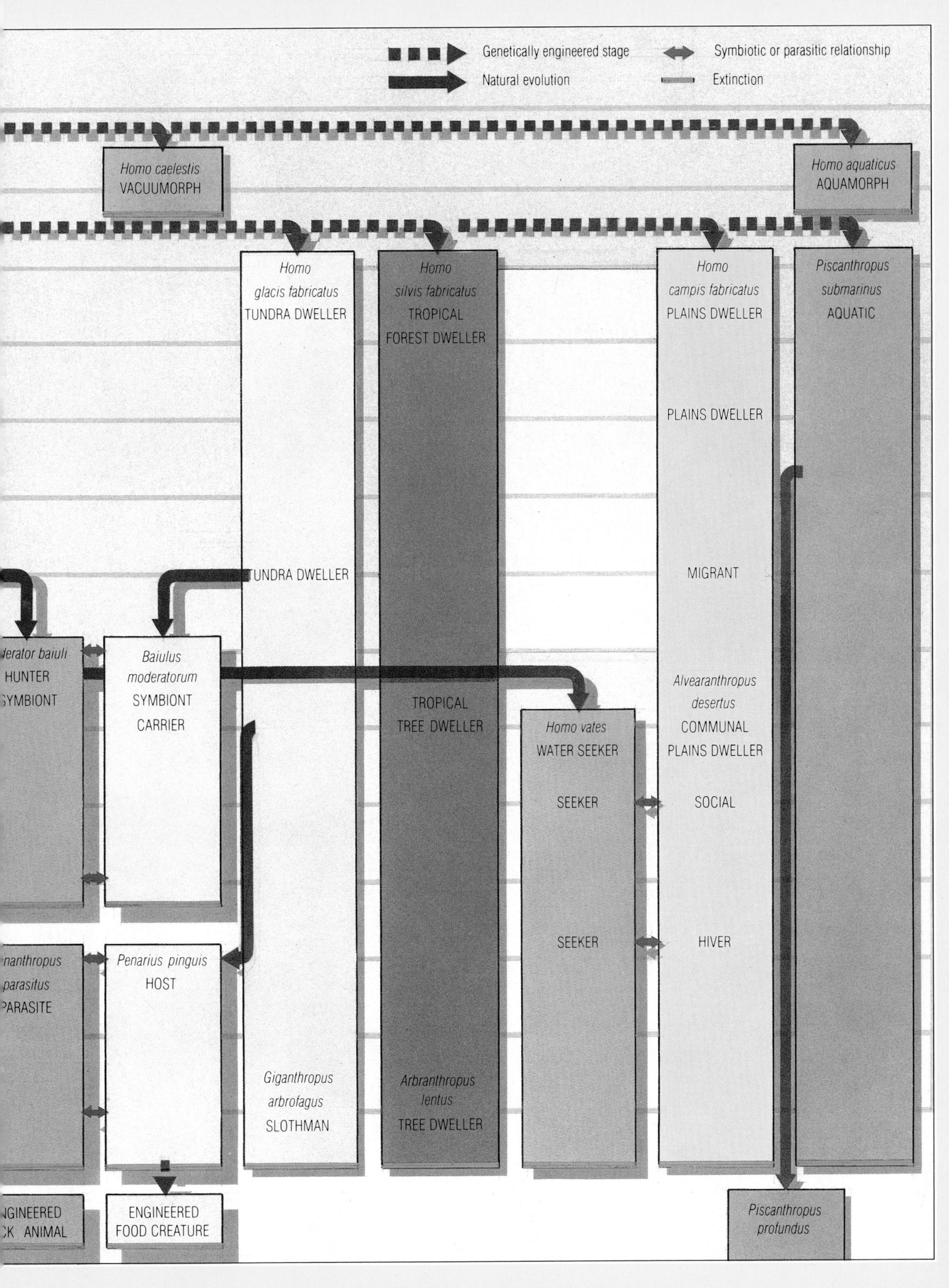

Genetically engineered stage
Symbiotic or parasitic relationship
Natural evolution
Extinction
Homo caelestis
VACUUMORPH
Homo aquaticus
AQUAMORPH
Homo
glacis fabricatus
TUNDRA DWELLER
Homo
silvis fabricatus
TROPICAL
FOREST DWELLER
Homo
campis fabricatus
PLAINS DWELLER
Piscanthropus
submarinus
AQUATIC
PLAINS DWELLER
TUNDRA DWELLER
MIGRANT
derator baiuli
HUNTER
SYMBIONT
Baiulus
moderatorum
SYMBIONT
CARRIER
TROPICAL
TREE DWELLER
Homo vates
WATER SEEKER
Alvearanthropus
desertus
COMMUNAL
PLAINS DWELLER
SEEKER
SOCIAL
SEEKER
HIVER
nanthropus
parasitus
PARASITE
Penarius pinguis
HOST
Giganthropus
arbrofagus
SLOTHMAN
Arbranthropus
lentus
TREE DWELLER
NGINEERED
CK ANIMAL
ENGINEERED
FOOD CREATURE
Piscanthropus
profundus

PART TWO:

MAN AFTER MAN

200 YEARS HENCE

PICCARBLICK THE AQUAMORPH

The grey-green of deep water is floored by a bed of rubble, sprouting wisps of red algae and sparse fan coral. Rusting steel hulks, caked with sponge and algal growth, jut up in incomprehensible shapes in the gloom. A few fish move slowly in the dark hollows, as the occasional scuffling crab raises brief clouds of sand and silt particles with its pointed feet.

Suddenly these few creatures dart for cover, as a much larger shape bounds its way slowly over the bottom. It is streamlined, as are all swimming animals, and its surface is smooth and rounded, all angles padded out by a thick layer of insulating blubber. The legs are somewhat frog-like, with webbed feet, but the webs continue up each side of the leg as far as the knee. The forelimbs are prehensile and adaptable, but for the moment are held tightly against the torso so as not to disrupt the streamlined shape. The creature gives off an air of deep sadness, but only because of the face, with its big dark eyes and an enormous lugubrious downturned mouth. The mouth funnels into a broad throat that connects to a wide belt of gills across the chest.

It ceases its movement and crouches on the bottom, looking upwards through water above it. Up there is a whole new world, a world that should not be strange since it is the world of the creature's immediate ancestors.

Its great-grandfather was a librarian, Jon Artur Blick, looking after and cataloguing the accumulated knowledge of centuries of human civilization. Its grandfather, Jon Blick Jr, was an artist, contributing to that civilization's culture. Its father, Jon Blick III, was an astro-physicist, adding to the information mankind could draw upon. Now Piccarblick is an aquamorph – a creature engineered to be part of a new frontier. This creature is human.

Piccarblick rises slowly towards the undulating silver ceiling that separates home from the hostile environment above. He rarely comes to the surface since he is not directly involved in trading with the land people. Whenever he does he is always uneasy, even though it was the environment of his parents. A flurry of bubbles arises about him as he ascends towards the surface. Controlling his ascent so that the pressure on his tissues is not released

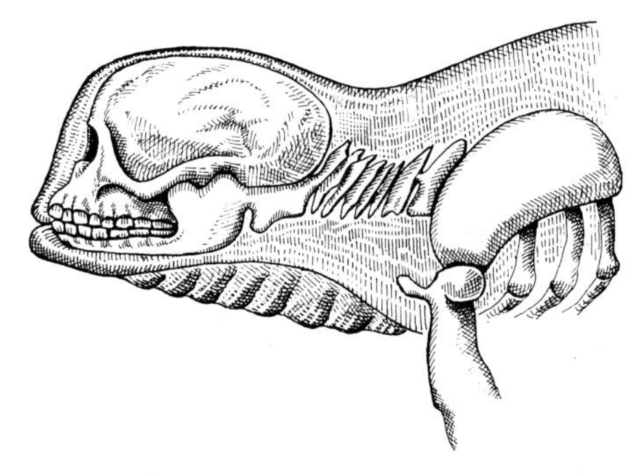

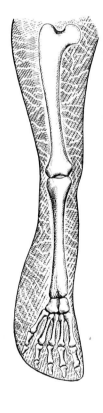

The skull is shaped and positioned so that a rounded head and short neck add to the streamlining.

The lower leg of Homo aquaticus *forms a powerful, well-muscled paddle, spread by the toes.*

Facial expression for the aquamorph is limited to basic responses. It relies on simple sounds to communicate.

200 YEARS HENCE

THE AQUAMORPH

Homo aquaticus

Fish-like and frog-like, the aquamorph is genetically adapted to live within a totally marine environment. Each physical feature – the streamlined body with the smooth skin and the insulating blubber layer, the gills on the chest, the paddles on the legs – was grown by the embryo. But this embryo was the result of manipulation of the sperm and egg cells. The chromosomal make-up was adjusted, creating genes that would produce features such as skin with a low drag fractor, and the whole organism was allowed to grow to its designed form.

Sliding easily towards the surface, the powerful aquamorph prepares to face brief contact with the hostile environment of its genetic ancestors. It does not envy the clumsy land-dwellers their damaged habitat.

too quickly, he bubbles up through the final few metres and breaks the oily, scum-laden surface.

Many of his family are already there. He can just make out their heads bobbing around him amid the floating rubbish. The sky, grey-white with an orange tinge of smog along the horizon, has an alien beauty about it – like the sparkling surface of the unpolluted Earth as seen by the first astronauts.

He looks towards land, but it is indistinct. His eyes will not function properly, because the difference between the refractive index of air and water is such that he cannot focus on anything above the surface. From his utility belt he takes air goggles and slips them over his head.

Now he can see clearly. The strip of rubble beach is backed by towering brown and black buildings of the land people. Down coast the buildings protrude from the sea, built on top of those already submerged, using the drowned hulks as their foundations and piles. The city he sees will not last for long, as the sea levels are continuing to rise and this area will also have to be abandoned.

Not, however, until the establishment has served its purpose. On the flat runway lies a narrow pointed cylinder, too distant for the details to be seen, but Piccarblick knows it from descriptions. Beneath the small wings at the rear lie the huge oxygen-compressing rocket engines that will heave the craft off the ground, through the successively thinner layers of atmosphere and eventually into orbit. There it will rendezvous with the starship, transfer its passengers and return to the runway.

The starship itself is complete and almost ready to go. All his life Piccarblick has been involved in its building. He and his family worked the great underwater deuterium mills that produced the fuel to power it, and farmed the continental shelf to sustain the land-dwellers and space-dwellers while they constructed it. Before long, fully manned and equipped, it will move out of Earth orbit, build up speed through the solar system and leave the regions of known space for ever. Its departure will mark the end of the work of Piccarblick's life. He and his fellow aquamorphs have toiled away, knowing that mankind's future may not lie on this dirty planet or in its polluted waters, but elsewhere in the cosmos.

Warning klaxons sound across the water. A flight of scavenging birds takes to the air from the beach as smoke bursts from the tail of the distant craft. After a seeming age the rumbling roar sweeps over the floating observers and slowly the vessel increases its speed along the runway and lifts itself into the air. Out over the sea towards the watchers it flies, rising as it goes. The sound builds up and, as the elongated shape hurtles overhead, the impact of the noise disorientates sensory organs more used to picking up water-borne signals. Then the ship is gone, leaving a lingering trail of smoke that slowly dissipates and adds its particles to the weight of atmospheric pollutants that have been building up for the past few centuries.

Piccarblick and his colleagues watch the ship go. Although excited by the sight, they remain silent because they cannot speak above the water. Quietly they turn over and dive back into the depths, where, as they descend, they can chatter freely to one another. They are home.

CRALYM THE VACUUMORPH

Cralym does not take after her mother, nor does she have her father's features. Both her parents were conventional unengineered humans, exactly like those that flourished and expanded throughout recorded history, reached their peak late in the twentieth century, and then declined under the weight of overpopulation, dwindling resources and hurtling environmental deterioration.

The genetic engineers took her ovum and his sperm, and altered their genetic make-up according to what would be required for survival away from the Earth's atmosphere, united the two and let their offspring develop in an extra-uterine environment on the orbiting laboratory 200 kilometres above the decaying Earth. The body matured and developed as a being able to live under conditions of weightlessness. All organs that had evolved to work in conjunction with gravity – legs and feet, hands with palms, sturdy backbone – were suppressed. The new legs and feet looked and worked more like arms and hands, and long fingers grew from muscular wrists; all these emerged from a compact spherical body designed to contain the pressures of the internal biology. Extra artificial organs that could not yet be developed by genetic manipulation were then inserted, such as the third lung used as a temporary oxygen store and the fourth lung used as a dump for carbon dioxide and other waste gases. The sealed-lens eyes and impermeable vacuum-proof outer skin, grown from tissue cultures in the orbiting biological vats, were later grafted on. The result was Cralym.

Throughout history animals were bred for particular purposes. Cattle were taken from the wild and mated with different strains, to produce varieties that developed more milk or more palatable meat. Selective breeding produced dogs with long legs that could hunt swift-footed animals, and dogs with long narrow bodies that could run down burrows and hunt subterranean animals. It worked. It was part of the influence that civilization had on the natural world.

When it came to adapting human beings in the same way, however, that was different. It implied a choice imposed by some individuals upon other individuals. It implied the wielding of a moral power over those who did not share that particular morality. It implied the deviation of human development from its natural course – a course perhaps decreed by a deity. It implied the making, not only of a body, but of a soul; and that soul would not have been acceptable in any of the faiths of the world. You could do all that to animals – but not to human beings. The concept was reviled by the word 'eugenics'.

Nevertheless, a time came when ethical considerations had to be compromised. If humanity were to survive, then it had to change. With the old system of selective breeding, the genetic material from one chosen individual was combined with that from another, in the hope that the desired attributes of each would appear in the offspring. It was a gamble. Genetic engineering was different. The precise function of each gene in the human system was now known, it was possible to manipulate it: to kill off a certain gene that produced an undesirable attribute, to add another that would emphasize a particular physical feature. Now beings could be produced to any specification.

Now aged 25, Cralym climbs along the outside of the starship's hull, gripping the struts and rungs with her toes and pedal thumbs. Her grip is now an entirely reflex action; very rarely has she lost her hold and drifted uncontrollably into the void. On such occasions she has been able to return by venting waste gases from her fourth lung and steering herself back to the ship. Someday the engineers will be able to develop some organ that will allow the vacuumorphs some effective locomotion through the vacuum itself.

With her pressure-sealed eyes tinted against the glare, she watches for the ferry to rise from the dazzling white and blue of the Earth below. She is unsure of its precise arrival time, but hopes that she will see it before she has to return to the interior of the ship. Sooner or later she will need to recharge her third lung with oxygen. At the moment she is quite relaxed, safely protected within her spherical exoskeleton from hard vacuum and cosmic rays – the environment for which she was developed. By custom Cralym is referred to as 'she', because of the original genetic make-up. The title, however, is a formality since she is neuter. Someday, perhaps, it will be possible for a heavily-engineered being to breed – but not yet.

It took 20 years to build the starship, and it will probably be only the first of many as eventually mankind, in one form or another, will spread out across the whole of the galaxy. The ship is shaped like two great conical spinning tops, fused nose to nose. The forward cone is the living chamber, a little world in itself that will have to be home to several hundred people for probably as many years. Around the waist is a ring of spherical propellant tanks, containing 30,000 tonnes of helium-3 scooped from the gases of Jupiter's atmosphere, and 20,000 tonnes of deuterium distilled from the Earth's oceans, all compressed into frozen pellets. When running, these pellets will be injected by electromagnetic gun into the aft cone – the reaction chamber – where they will be compressed into a fusion reaction by high-power electron beams. Magnetic fields will direct the continuous blast rearwards and the entire vessel will move out into unknown space, accelerating continually as it goes, eventually reaching about 15 per cent of the speed of light. The people who go with it will never return.

That does not include Cralym, who would not have been sorry to leave Earth orbit. She has never set foot or hand on Earth itself, nor has she ever had any wish to, but she would have liked to travel to another planet, another system around another star. She could never have survived the journey, however, as she was designed for living under the conditions of zero gravity in space. The starship, flying under a constant acceleration, will generate its own gravity, and allow non-engineered humans to live without problems. It will be crewed by the non-engineered, but genetic engineers will be amongst the passengers. Who knows what conditions they will meet, and need to adapt to, on a planet in a distant star system?

A glint of light is caught and reflected in her sealed lens. The ferry will soon begin its docking manoeuvre as it drifts towards the starship. Cralym and her fellows clamber along towards the port to watch.

JIMEZ SMOOT THE SPACE TRAVELLER

Jimez Smoot can breathe again. The acceleration of the take-off and the lift to orbit had squashed the breath out of him. Now, as the ferry coasts into free-fall, he and his

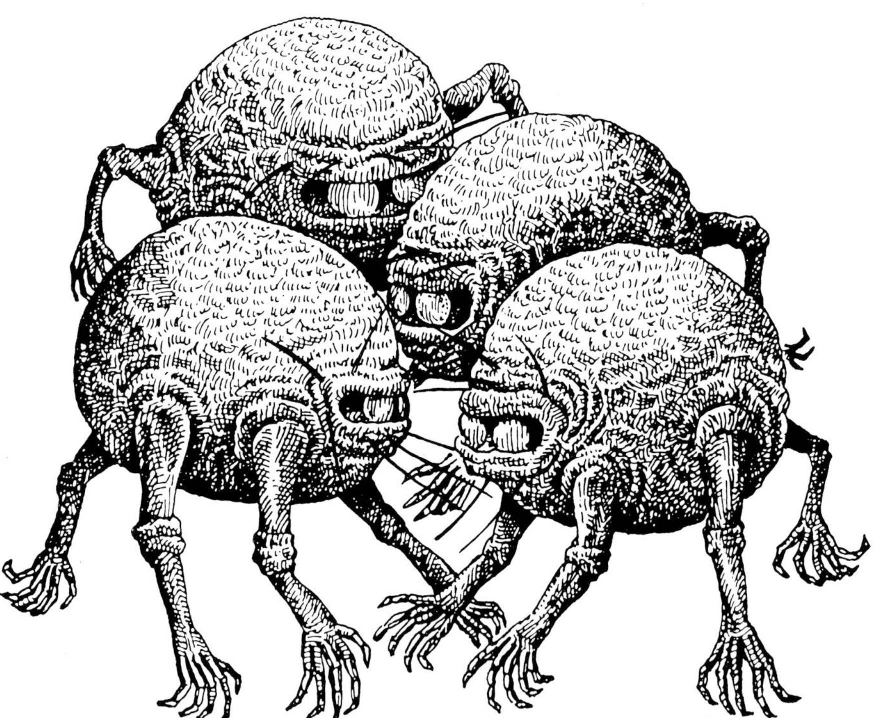

Without sound, communication in space must be by touch, using their sensitive whiskers.

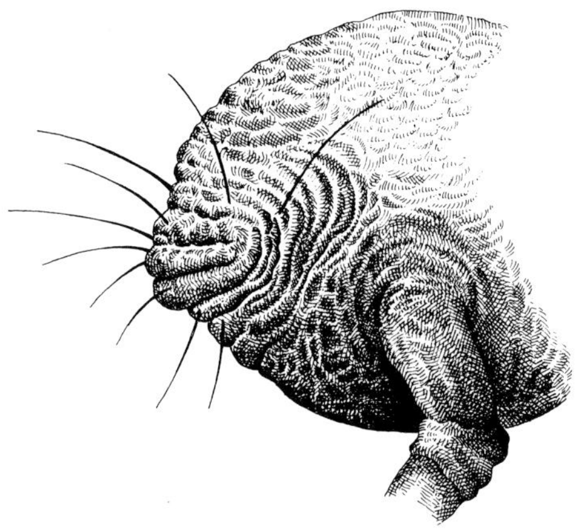

Heavy lids shield the eyes against solar wind, while a sealed lens protects them from the vacuum.

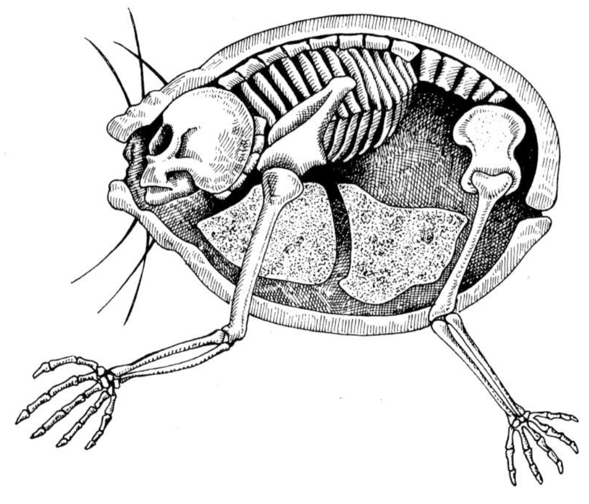

The spherical shape and the hard outskin keep in the body pressure, and contain the additional organs.

200 YEARS HENCE

THE VACUUMORPH

Homo caelestis

Protected from the harsh glare of earthlight, tinted eyes peer deep into the darkness of space. If humanity has a future, it is there that potential might be.

The ultimate triumph of the genetic engineer. The product of grafting, surgery and cell manipulation, the vacuumorph can live and work in the free-fall of orbit and the airless void of space. The basic human reproductive cells were manipulated to produce the necessary shape, but extra organs had to be grafted on, including a hard impermeable outskin grown from tissue culture. The result, however, is sterile. *Homo caelestis* has a limited life and no future of its own. The vacuumorph cannot breed and would not survive the rigours of gravity.

200 YEARS HENCE

JIMEZ SMOOT

Homo sapiens sapiens

Only the most perfect human specimens are being sent to colonize the stars. Perfection, in this case, is clearly defined. Each colonist is carefully selected to ensure that his or her biological make-up is as flaw-free and reliable as possible. Space will be its own habitat. And later the surgeons will require the best raw material available when it becomes necessary to build new living creatures to fit those unknown environments to be found far beyond the solar system.

Physically fit, psychologically stable and rigorously trained, Jimez Smoot is the raw material for one of his world's most dramatic and desperate experiments – the epic conquest of space.

man
aquamorph
vacuumorph

Homo sapiens sapiens
Homo aquaticus
Homo caelestis
Homo sapiens machinadiumentum
Homo virgultis fabricatus
Homo glacis fabricatus
Homo silvis fabricatus
Homo campis fabricatus
Piscanthropus submarinus
Homo sapiens accessiomembrum
Homo mensproavodorum
Speluncanthropus
Moderator baiuli
Baiulus moderatorum
Homo dormitor
Homo vates
Alvearanthropus desertus
Homo nanus
Nananthropus parasitus
Penarius pinguis
Piscator longidigitus
Formifossor angustus
Acudens ferox
Harenanthropus longipis
Giganthropus arbrofagus
Arbranthropus lentus
Piscanthropus profundus

fellow passengers are released from the gravitational grip of his planet – for ever. He was trained for the accelerations and the weightlessness, of course, in the acclimatization camp back home, but no amount of training on simulators could have prepared him for the actual power and terror of the real thing.

Back home? Yes, he will probably think of Earth as 'back home' for the rest of his life, although he will certainly never go back there. He was taken from his community and subjected to years of rigorous training for survival in a small group, the members of which were drilled to accept, psychologically, that they would be going on a journey from which they would never return, and which could well end in failure and waste rather than the founding of a new civilization. He and his colleagues will abandon everything that they have ever known.

They are the lucky ones.

No amount of scientific application could have stopped the deterioration of the Earth's environment. No amount of moral guidance or medical technology could have slowed the suicidal birth-rate. No amount of exercise of the new-found science of genetic manipulation could induce crops to produce enough food in the right areas to feed all. No amount of political co-operation ensure a fair distribution of what was available. The most influential cultures relied entirely on technology; and stable conditions were necessary for their technologies to flourish. As their systems crumbled, one by one, the less technological peoples spread to take their place; but inherited the economic systems left by their predecessors, and quickly adopted and rebuilt the highly technological way of life, with all the disadvantages and disasters that implied. Desperate and amoral, but practical, choices were made. New beings were genetically engineered to fit into the uninhabitable environments, so that new areas could be exploited for the good of the whole.

Yet industry still generated its waste products. The carbon dioxide level in the atmosphere rose and the world sweltered under an accelerating greenhouse effect. The icecaps began to melt and the sea levels worldwide rose. Already most of the great cities of the world, those built on deltas, estuaries and low-lying coasts, were flooded and uninhabitable. Temperatures at the equator became intolerably high and populations migrated to cooler latitudes, abandoning the stripped rainforests and the desertified grasslands.

All this change came about in a mere two centuries; but those two centuries were the culmination of 4000 years of burgeoning civilization. Yet, on the time scale of the planet, 200 years, even 4000 years, were hardly noticeable in the context of the 4500 million years that the planet had been in existence.

We will do it better next time, Jimez Smoot muses as he looks out of the porthole. The dislocated Earth systems are invisible from up here – just the odd smear of brown or yellow showing where smoke is particularly concentrated. Across the cabin, past his companions and through the opposite porthole, he can see the ungainly shape of the starship as they approach it. The airlock on the payload module, surrounded by the tiny figures of the vacuumorphs, beckons invitingly. This will be their new home. Neither he nor his children may be the ones to begin the new civilization on a new planet; but eventually the ship will reach somewhere habitable and one of his descendants will be part of the new system of humanity, which will benefit from the lessons learned from the mistakes of the old.

KYSHU KRISTAAN THE SQUATTY

The moving lights in the sky, glimpsed through a brief break in the clouds, mean nothing to Kyshu Kristaan. He had to work hard to steal the meagre portion of food he carries wrapped in his rags, and he cannot let his attention waver from his journey for an instant. Now he has to step quietly over the sleeping huddles that lie in the dry alleyway. Should anyone wake up and sense what he is carrying then he would have to fight again, and this time, exhausted from his last skirmish, he would not win.

His wife and seven children lie, hungry, in an old culvert beyond the flooded thoroughfare. They have not eaten for three days. That was the last time the reliefer flew over and dropped food on the drowning city for the squatties. The relief drops were becoming more and more irregular, and Kyshu Kristaan knows why. Inland government does not care about its people; all it cares about is producing monsters: monsters in the seas, monsters in space. As for the weak mockeries of human beings, the hand-fed who have never had to fight for their food in their lives, the soft ones that they are sending to the stars... Kyshu does not even want to think about them.

At the entrance of the alley the road slopes down into the oily water. With his sharp eyes Kyshu Kristaan can make out the tangled tidemark of discarded artefacts, empty containers and decaying bodies that lies along the edge of the water. Tide is out, he registers, so he can wade across. His

200 YEARS HENCE

KYSHU KRISTAAN

Homo sapiens sapiens

In the steamy shell of the once-great cities life is brutal and short. Disease is rampant. Starving hordes, the future beyond their control, squat in the ruins of collapsed civilizations. Electricity and water supplies have failed. Food is scarce and only the strongest and most determined can survive – and then only for a while.

As food becomes scarce, order becomes a luxury. Civilization has been replaced by a society on the edges of chaos. Boundaries are clearly defined, and family groups fight to defend their territories.

brother died because he did not have such good eyesight, and would not have noticed whether the tide were in or out. He had not seen the man who lay in wait for him in the darkness as he returned home with his food. Kyshu Kristaan hopes that his own children will inherit his sharp eyesight; Sem Kristaan died before he could father any children and so his weak eyesight died with him.

Chest deep, Kyshu Kristaan crosses the flooded thoroughfare, holding his precious bundle high so that it will not become any more tainted. Slowly, making sure that no ripple breaks and makes a warning noise, he rises from the water at the other side. Yet, as he pulls himself from the water the silent night shatters into a cacophony of yells and shrieks, bursting from the direction of his sewer home.

No! – he despairs. Not more fighting tonight. Other people do not live like this. The inland people, the ones that have things to do, do not live like this. There is plenty of food for them and they do not have to fight.

Now his wife Seralia is yelling and his children are screaming, and he drops his bundle and runs to help. Seralia is at the culvert mouth, flaying about her with a steel bar. There are still figures at her feet and others grabbing her by the arms and legs. Kyshu surprises them all. Seizing the hair of two of the attackers, he cracks their heads together with a force that must kill them; then he turns and fells a third with a blow of the edge of his hand. The others disappear into the darkness, as do the shadowy watchers.

Then, while Seralia calms the frightened children, Kyshu gropes back to where he dropped his bundle, but it has gone, stolen by one of the dark watchers. An exhausted depression creeps over him. Must it always be like this?

Seralia calls to him for help, as she drags the bodies over to the water to deposit them (no point in fouling the entrance of their own home). It is said that the squatties in the next city eat the victims of a fight like this. That would be practical, but inhuman, although Kyshu could believe anything of them. Their attackers appear to be from across the water, probably on their way back from a raid, and their ragged pouches are full of stolen food: fair compensation for the food that Kyshu has lost, so he does not feel so bad now. Six men have been killed, which is a good total for such a skirmish.

He and Seralia can feel proud of the night's work. There may be monsters in the sea, and monsters in space, and the weaklings can be sent to the stars; but it is here, in the squats of the drowning cities, that true men and women are still to be found.

300 YEARS HENCE

HARON SOLTO AND HIS MECHANICAL CRADLE

Haron Solto opens his eyes to the soothing light display on his ceiling. Gentle scents exude from the walls, and a hot drink appears instantly in the alcove by his hand.

He speaks a word that would have been meaningless to twentieth-century people, but it banishes the light display and surrounds him with bland walls. He lifts his one good hand and takes the beaker of drink. The cakes of synthetic protein that constitute the first meal of his day immediately appear behind it. The tastes are delicious – synthetically produced, but delicious.

Having eaten his first meal, he goes for his hygiene session. By moving one finger of his wizened left hand, contact is made and his cradle hums and moves his body towards the ablutoir. The whole contrivance, mechanical and biological, travels on its magnetic levitation motors across the chamber towards the arch that houses the ultrasonic cleansers. All the switches and contacts for his cradle are within reach of the little fingers of his left hand. A panel above them shows him instantly that all his life-support systems are working. The cultured kidneys distil, the synthetic liver produces the chemicals that help the digestion of his first meal, the external lungs circulate enough purified air, the metal heart pumps blood through the biological part of his being, and all the motors, relays and servo-mechanisms that provide the mobility of the unit function properly.

Time for his day's work.

Away below him, in tunnels deep within the Earth, lie the protein factories. Fully automatic and never seen by humans (except for the handlers who repair and maintain them), the factories use the power generated by the mountainside full of solar cells to turn raw carbon dioxide and oxygen from the air, and water from the reservoirs, into edible carbohydrates. Elsewhere lie the machines that simulate the biological processes that produce edible protein, and still other factories house vats that produce the flavours and textures that are added later to the world's food supply, and help to change food from a mere nutritional necessity to an art form.

On his personal display Haron Solto sees the figures for

300 YEARS HENCE

HITEK

Homo sapiens machinadiumentum

When biological organs consistently fail, substitutes must be developed. The more vital the failed organs that cease to work, the higher the technical back-up needed. Scientists are already working to produce tissue-based replacements.

As long as the brain functions, it is worth keeping it alive – even if the body has deteriorated.

the day's world requirements, its preferred geographical distribution, the output from the various factories, and the flow of the transportation systems. With a practised eye, he reads the graphs and evaluates the estimates; then three quick presses of a button and the day's production is in balance. He can rest.

The motor units of his cradle lift him away from his workstation. Today he will contemplate his sculpture collection, which always gives him peace of mind. He sweeps across the room to where the three-dimensional images are housed, but beneath the healthy hum of the levitation motors there is another sound – a hissing and grinding noise. His forward motion ceases abruptly, an edge of his cradle tilting and scraping the floor.

Panic! No, don't panic: it can all be controlled. Punching a button, he injects the right amount of sedative into his system to regain calm. It was a minor malfunction of his locomotor system, nothing more. Summon up one of the handlers, immediately.

After a short time, a time when Haron Solto is beset by thoughts generated by helplessness and indecision, the handler appears in the external door. He is a primitive, like Haron Solto's ancestors must have been, and obviously male. He walks without mechanical aid, and his body is symmetrical, with two arms and two legs. Like all handlers, he will have been taken from the outside ruins. Their versatility makes them useful, and they are willing to perform distasteful functions in return for food and comfort. This creature has few mechanical appliances, but his body is covered with an insulating clothing, and he carries a bag of instruments around his shoulder. Solto tries to close his mind to the disgust he feels; but has to acknowledge that these people are necessary.

With a few words, in a dialect that Haron Solto can hardly understand, the handler diagnoses the problem and sets to work. A panel of the cradle is removed, then tools and artefacts are brought out of the handler's bag and fixed into the mysterious innards of the machine. It all takes place out of Haron Solto's sight. The inside of the cradle is something he has never seen and has no wish to see. All he perceives is the hairy top of the handler's head as he bows over his work, making a humming noise with his lips and teeth, a noise that Haron Solto surmises passes for music.

The experience is too much. Haron Solto blasts some narcotic into his vein and drifts into a more pleasant substitute world.

He is roused by a loud slam, as the panel in his cradle is rammed home. The handler says two brief words to him. The first denotes work done, and the second is a mode of address, basically respectful but which, Haron Solto suspects, has now become a term of amusement and mild mockery amongst the handlers.

Haron Solto dismisses the man, having first endorsed his identity chip to say that the work has been done.

Haron Solto is alone once more, fully functioning, and can continue his day's reverie. Someday humanity will not need these grotesque throwbacks to primitive man. There will be a better method than the present mechanical contrivances: a system that lives, grows and repairs itself. That is for the future, however, and someone else will have to develop it.

GREERATH HULM AND THE FUTURE

Humanity has a potential which cannot be bound by mere machines. There must be a better way forward.

These are thoughts that have beset Greerath Hulm ever since she witnessed the last failure of the local food generator. It was a terrible time during which the handlers fought amongst themselves. On one side, the disciplined faction struggled to repair the breakage; on the other, those whose food supplies had been cut off first were trying to break into the machinery to feed on the raw materials. Order was restored, but only through massacre.

What had human beings come to now? Wizened bodies encased in machines, kept alive by mechanical contrivance and synthetically-grown organs.

Once, a long long time ago, humanity developed through the process of evolution. With the coming of intelligence and civilization, this natural process was swept away. Medical science developed, and those that would have died off were now able to survive and reproduce. As a result for man the directive force of evolution – the process of natural selection – was eliminated. In consequence, the species deteriorated. Unhealthy changes that would have been wiped out now survived and were spread. The genetic stock became weaker as populations became bigger. This did not matter, because medical science was always there to sustain life. No matter how degenerate a human body became there were always the technological systems to keep it alive.

The result was certainly a triumph over the raw wildness of nature, but there *must* be a better way. Machines keep breaking down and the food and drug supplies are constantly disrupted. Synthetic organs must hold the key.

If they improve, muses Greerath, that would put her and many like her out of work (she controls the manufacturing process for a series of synthetic enzymes and stimulants that benefit humans the world over). That might not be a bad thing. She would like to devote more of her time to listening to music, looking at art, and wallowing in the newly-developing medium of hypnotic-involvement-drama.

Then, with a start, she remembers two friends who recently retired from work to do just that – and both of them switched off their life-supports after a few days. Probably their stimulant-mix was wrong – something that will not happen to Greerath; after all, she is in the business.

Genetic engineering must be the future, though. Humans have already dabbled in it during the last century, when it produced beings that could live in space. That was specifically for work on the star-colony project; so, as always in history, a specific emergency or a specific goal fuelled a burst of technological development. In the past it was always warfare that provided the emergency. The technology usually involved the development of more sophisticated weaponry. Then, as ever, once the emergency passed and the goal was attained, the newly-developed technology fell fallow. Now that the star-colony project has come to an end, and the last of the 37 ships has been dispatched, there are no more space children. Those vacuumorphs were never perfect; they were not so much bred as built up from pieces grown synthetically, and there was never a possibility that they would reproduce. The aquamorphs, the humans engineered to live in the sea, are still there, though, living in the warmer waters of the ocean. A veritable underwater civilization is developing.

A burst of sunlight from behind the clouds, slanting down the fissures between the tall buildings, cut to geometric dapples by the supporting girderwork, and discoloured by the translucent filters of Greerath's habitat, creeps into her living unit and brings her out of her daydream. Her day's work is almost over, and she has hardly done a thing. Once, she thinks, mankind was ruled by the sun: when it rose people woke up and started their day, and when it set they slept. Now nobody could care if the sun were there or not – as long as it powered the solar cells, and kept the ocean currents churning away and driving the submerged energy units.

Out there, where people no longer go, there are wild spaces on the planet. For a while these were poisoned. Now all that has changed. The big animals have gone, all right, but the plants have re-established themselves. Steamy tropical forests are growing again along the equator, and grasslands lie in belts to the north and south. Further north and south are the spacious deserts that, because of the natural pattern of circulation of the wind and moisture, will never be fertile. Beyond these there are deciduous and coniferous forests, then towards the north and south poles lie the cold tundra regions and the icecaps.

Greerath knows of all these things from the information banks, but the subjects with which she is most familiar are found in old recordings. The tropical forests she now visualizes were full of monkeys, tapirs, anteaters, snakes, sloths, apes, jaguars, humming birds, toucans and eagles. The grasslands were alive with herds of zebra, elephant, antelope, giraffe, and pursued by lions, cheetahs and hyenas. The deciduous and coniferous forests had deer, beavers, squirrels, badgers, wolves and lynx. The tundra supported reindeer, musk ox and foxes. She knows that now these animals are all gone, and are as relevant to the modern world as are the dinosaurs, the moas and the mammoths. Today these habitats are open and silent, with only the smallest rodents and birds living there, along with insects and other invertebrates.

Surely out here should be the future of mankind? If so, a renewed campaign of genetic engineering could be the means of reaching it.

HUEH CHUUM AND HIS LOVE

It is probably the most dangerous and most exciting time of his life. Hueh Chuum is slowly and purposefully disconnecting himself from his cradle. For a few brief minutes he will be isolated from the things that keep him alive – but it will be worth it.

He has been preparing for months. Gradually his physicians have been turning off his libido suppressant. He has been trained thoroughly as to when to switch off this device and that organ. Those that are fundamentally necessary to his continued existence are connected to trailing cables and tubes – vulnerable but necessary for the essential few minutes. He is luckier than most: his heart is his own.

It is almost time. His sensors tell him that Bearnida, his love, is outside the door. He has seen her before, but only on screens and holograms, and was first attracted to her by the way that she had decorated her cradle. He realized that this attraction was not as superficial as it seemed. Her artistic taste showed that, deep down, she was similar, and that they would make a good mating pair. She approved, as did all her colleagues, physicians and relatives.

The environmental lights dim to a soft hue, and the ambient odours and music produce a gentle and seductive atmosphere. The access slides open and Bearnida's cradle wafts in.

He is seeing her for the first time without the help of mechanical media. Only her face, of course, is visible, and it looks a little smaller than he expected. The decorations on her cradle are bright and flamboyant, as befit the occasion. Inside the mechanisms, he knows she has switched off her life-supports for the short time necessary. She smiles at him, and he returns the smile – the first purely personal communication he has had with anybody.

The cradles drift together and their touching panels open. The lights go out – for who wants to see the wizened deformed body of a naked human, however much in love they are? Hydraulic arms, supporting the little bodies in their harnesses, swing out until they meet. . . .

It is much, much later. Hueh Chuum's shock is beginning to wear off and grief is settling in, but that can be dealt with by suitable injections. He is back in his cradle where he is safe. He is never coming out again as long as he lives. Never!

He thought that he and Bearnida were well-matched, not only mentally and emotionally, but physically too. Like him, she had her own heart; but hers was not nearly as strong as his, and the strain of mating was too much.

He can console himself that he is not alone, as only about 10 per cent of matings these days are successful. If this goes on, the human species will dwindle and die out.

AQUATICS

The sea waves, blasted by a south-western gale, curdle and foam in cold blue slopes that march remorselessly across the desolate surface of the northern ocean. From the lead-grey sky the chill rain hisses down in the icy green hollows and is lost in the streaming foam of the crests. The sea surface is not a welcoming place.

Below the screaming, tumbling chaos of the surface and in the top few feet of the ocean water the gale is silenced, the waves suppressed to a gentle to-and-fro motion. Further down, the movement becomes weaker and weaker until it dies away completely. This is the world of the fish – and of the creatures that have abandoned their life on land to accept their ancestral home in the great oceans of the world. To some extent the sea otters did this, with their sinuous bodies and their webbed feet; the seals and walruses did it more efficiently with their streamlining and their flippers; but the now-extinct dolphins and the great whales did it to perfection, even adopting the fish-shape of their forebears.

Now humans have done it too.

In the green half-light below the ocean's turbulence they swim. An unaccustomed eye might have taken them for dolphins, moving and turning, dashing away in a sudden streak, hanging for a while motionless.

They cannot breathe air, these creatures of the ocean. Instead they circulate the seawater through their mouths and pectoral gills, extracting the oxygen as it goes. They also feed constantly, filtering plankton through the same gills and transferring it to the digestive system. Now and again they take a fish – turning and streaking after it with a twist of the tailfin, a balance of the arms and a quick bite.

The tailfin is all that is left of the human legs. In embryo, the limb buds grow together and fuse into one organ. The hip girdle does not develop and the limb bones become almost an extension of the backbone. The phalanges of the toes spread and shape themselves into a network that supports the powerful diamond-shaped fin. The hands retain their human structure, but the arm has flattened and become modified into a balancing and stabilizing organ.

The development was started a century ago as part of the star-colony project, but the creatures developed were only partially successful. Later the engineering laboratories, in a last bid to produce something permanent before being closed down, refined the design and produced a truly aquatic human being; and (their final triumph) the genetic changes that they produced were actually hereditary. Yes, these newly-developed creatures were fertile, and produced viable offspring.

The process really started way back in the early days of civilization when man's quest to possess all the things of the world took him to the water. He invented mechanical devices that enabled him to take his air down into the sea with him and to breathe it at a workable pressure. Implements strapped to his body allowed him to see underwater and to swim with powerful leg strokes. As time went on great communities, rather like island cities, were established on the sea bed. The sediment-choked ruins of these still litter the continental shelves. When genetic engineer-

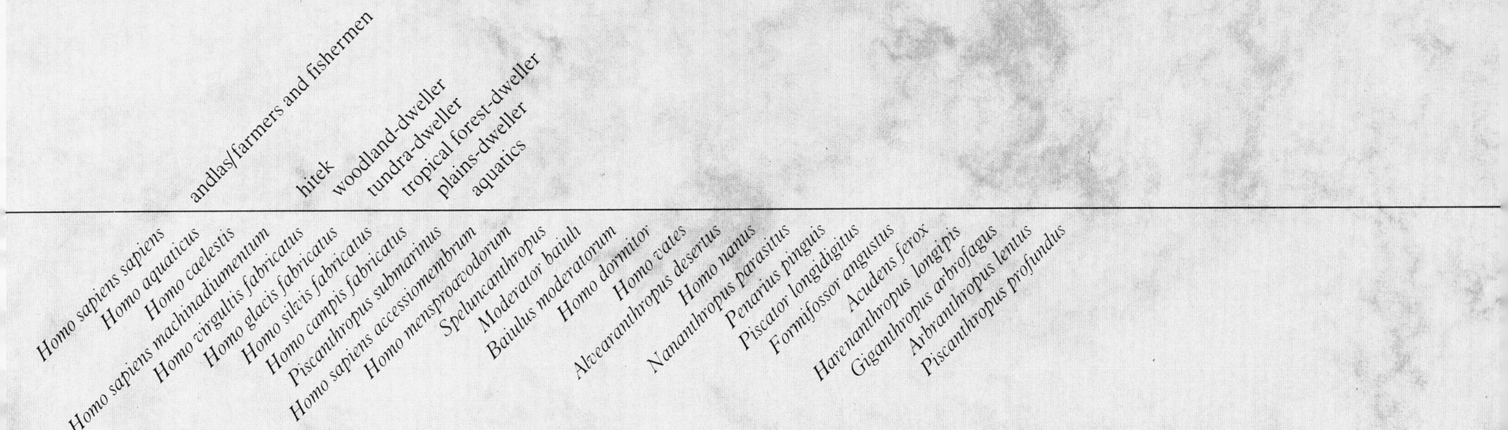

andlas/farmers and fishermen
hitek
woodland-dweller
tundra-dweller
tropical forest-dweller
plains-dweller
aquatics

Homo sapiens sapiens
Homo aquaticus
Homo caelestis
Homo sapiens machinadiumentum
Homo virgultis fabricatus
Homo glacis fabricatus
Homo silvis fabricatus
Homo campis fabricatus
Piscanthropus submarinus
Homo sapiens accessiomembrum
Homo menspropavodorum
Speluncanthropus
Moderator baiuli
Baiulus moderatorum
Homo dormitor
Homo vates
Alecaranthropus desertus
Homo nanus
Nananthropus parasitus
Penarius pinguis
Piscator longidigitus
Formifossor angustus
Acudens ferox
Harenanthropus longipis
Giganthropus arbrofagus
Arbranthropus lentus
Piscanthropus profundus

ing was developed, gills could be cultivated from raw tissue and grafted onto the human body, enabling humans to breathe like fish. This was still clumsy and imprecise compared with the later engineering of a creature with no need of cities or artificial swimming and breathing devices.

What swims here is merely the surface race of the creature. In the blackness below, hundreds of fathoms down, others exist, rarely seen by any but their own kind, and even then they are not strictly 'seen'. In the blackness they can only feel their way about and communicate with one another by a kind of echolocation. These creatures are sluggish and inactive. There is little food at these depths and they must conserve what energy they have.

Since the aquatics rarely meet any other form of human, there is no enmity between them and any other group.

A female suckling a wriggling youngster undulates gracefully towards a group of males who are chasing fish. She speaks. The 'voice' is a rattling sound, produced from clicks in the relict windpipe in the neck. The young males clatter their reply and swim off in what seems to be a random three-dimensional pattern. Suddenly the fish with which they were sporting congregate in a mass in front of the female's head, herded there by the precisely co-ordinated movements of the males. A quick flick and a snap, and she has swallowed one – the rest scattering into the green murk. She clucks her thanks to the males and swims sedately away. To look at, one would think that these are creatures that had existed in this environment since the world was young. It is only the face – a grotesque parody of the human face, with big bulging eyes, tiny degenerate nose and downturned mouth – that shows it to be derived from a human being.

500 YEARS HENCE

GRAM THE ENGINEERED PLAINS-DWELLER

Gram stands shivering on the dusty plain, not shivering with cold but with apprehension. The spiky grass round about is familiar enough; he has been brought up on a diet of it since he was born, ten years ago. During that ten years, though, all the grass he knew had been grown in the habitat module. He was brought up and cosseted by Family, a group of creatures that saw to his every need and trained him for life outside.

Only in the last two years did he realize that he was not like the people of Family. He was not encased in metallic outer skins, he did not glide along the floor and cables and tubes did not spiral out of him, connecting him to glass and plastic devices – and his face! The faces were the only parts of Family that he could see directly, and his was nothing like theirs.

Now he is on his own and he knows it. Family cannot live out here, on the grassy plains, so they are all congregated together in the flying machine behind him. All this landscape before him is to be his.

Delicately he steps away from the flying module. Beneath his tapering foot the fibrous soil feels strange – not quite like the soil in the habitat. He can feel the eyes of Family on him, as he wades into the sharp waving grass, scanning him closely, as he knew they would. Not only are they watching him directly, but the little instruments that are strapped to various parts of his body are sending back signals, telling them how he is performing.

He knows what he is supposed to do; he has been trained for long enough. As in the habitat, he reaches with his long arm and long hand and grasps a bunch of grass. The calloused cutting edge of his hand shears through the stems and leaves with a twisting motion, and he thrusts the bunch into his mouth and begins chewing. His big teeth grind into the stringy plant material, crushing it to a pulp and disrupting the fibres. He can feel the toughness, and knows that the wear on his teeth will be immense. He also knows that once a tooth is worn out another one will grow to replace it, and this will happen for the rest of his life, another thing that makes him different from the members of Family. He swallows the wad of grass, and down it tumbles into his voluminous stomach where it is met by specially-engineered bacteria that complete the digestion.

He scythes off another handful and eats it. This is working all right, he thinks, and hopes that Family think so too. He looks up to the horizon, a vast distance away. So this is to be his new home.

With sudden joy, Gram bounds away towards a clump of low bushes. He could be happy here, no matter what Family think. Suddenly he does not care what Family think: this is not their world – it is his.

Then in a first and final gesture of defiance he rips off the instruments that are strapped to his body and flings them away into the dusty grass.

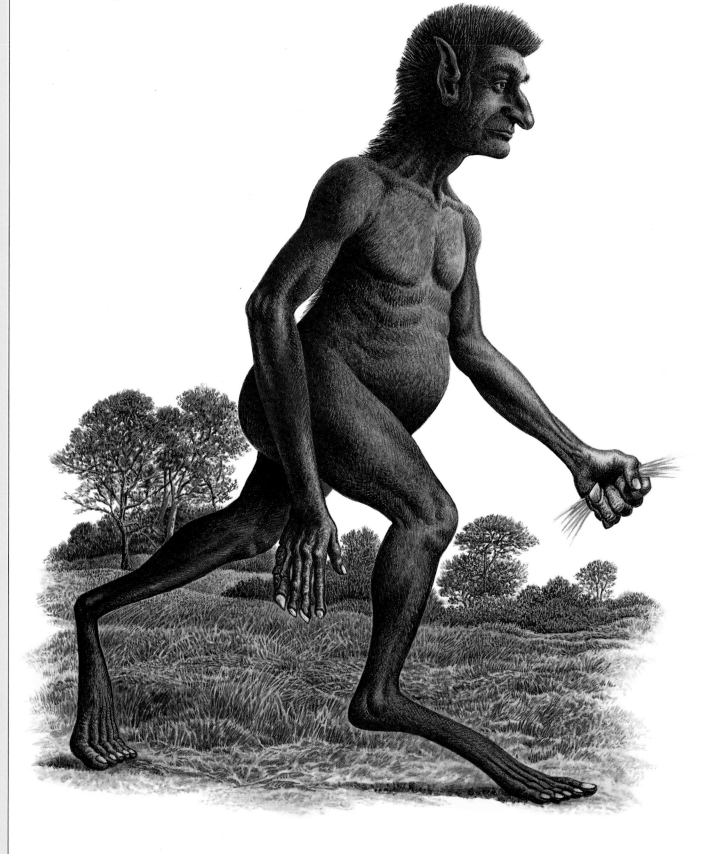

Harsh sunlight beats down on the plains-dweller's dark skin as he runs effortlessly through the dusty grasslands. Vegetation is tough and will also be sparse during the seasons of drought.

PLAINS-DWELLER

Homo campis fabricatus

A human engineered to live on open grasslands needs the adaptations of a grass-eating mammal. For the plains-dweller these include massive teeth that are replaced if they wear out chewing tough silica-rich grasses and, more importantly, a specialized stomach within the bloated abdomen containing engineered bacteria that can break down cellulose – a substance not normally digestible by the human frame. Cutting edges on the hands help to scythe the thick grass while the long legs enable the creature to move swiftly over the open landscape.

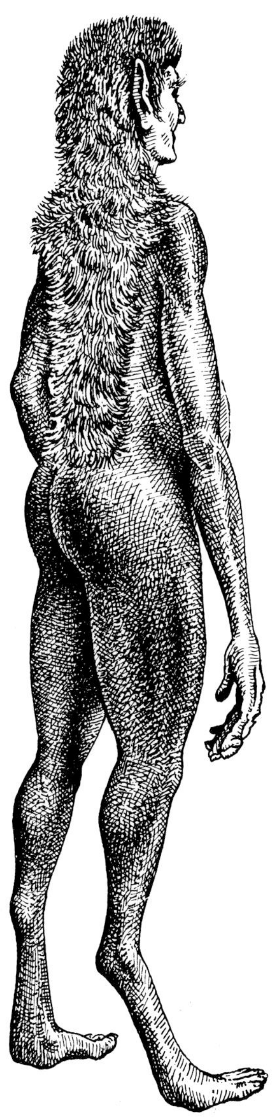

Blade-like callouses provide the plains-dweller with some degree of protection, as well as cutting through tough stems.

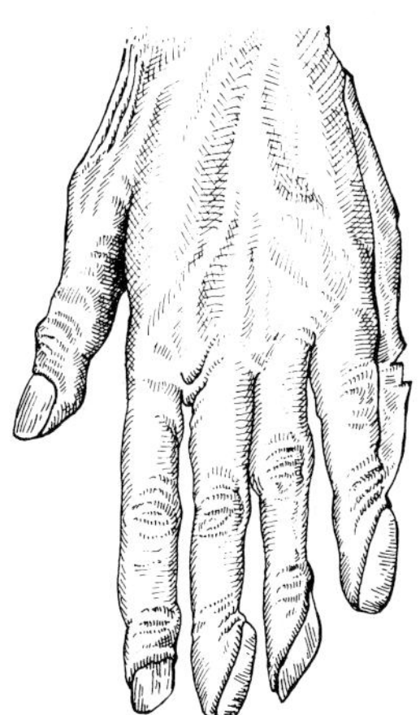

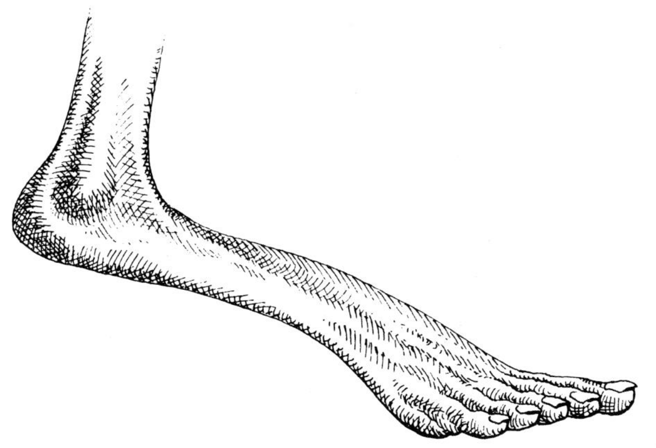

The dark skin and mane of hair across the shoulders and running down the back protects the grassland-dweller from ceaseless sunlight. The long feet have become an extension of the legs, adding to his speed.

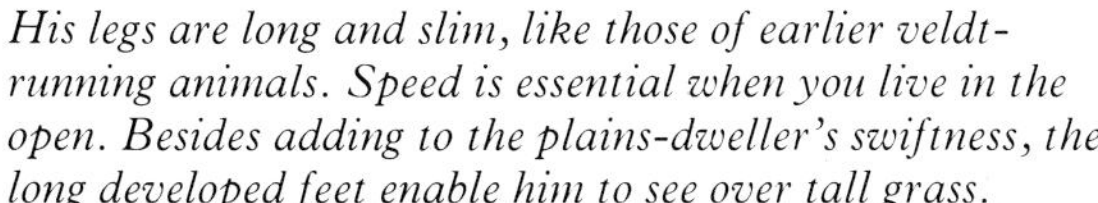

His legs are long and slim, like those of earlier veldt-running animals. Speed is essential when you live in the open. Besides adding to the plains-dweller's swiftness, the long developed feet enable him to see over tall grass.

KULE TAARAN AND THE ENGINEERED FOREST-DWELLER

Kule Taaran looks down at the huge oval shadow of the flyer falling on the top of the tropical rainclouds, surrounded by a rainbow ring of spectral colours. As the vessel descends, the clouds clear away below it and the vast stretch of green forest reaches out as an unbroken carpet with dark rivers winding across it. The flyer's shadow on the treetops is now fuzzy and unclear, but soon it comes into focus and the edges become sharper as it descends. Now individual trees can be seen, and with an uneven crunch the vast vessel settles amongst the broken branches and boughs.

Kule Taaran looks around him at the rainforest. It is not as it once was. A few centuries ago the first rainforest was all destroyed, as a burgeoning population of humans spread into it and removed it, clearing it away to make room for them to grow food. It was a disaster, not only removing the entire forest and its animal life from the face of the Earth but also producing subtle changes to the climate the world over. Such problems are all past now that there are more efficient ways of producing food. The forests have returned, but not in their old state. The forest soil which had taken millions of years to build up was nearly all washed away in the bad times, so the trees that have repopulated the area are not the magnificent trees of old. They are scrubby and hardy, adapted to find a roothold in what soil there is left; but the hot climate and the constant rain has made them grow prolifically.

No big animals exist, though. With the great trees of old went the monkeys, apes, jaguars, parrots, toucans, tapirs, squirrels, opossums, okapis and bongos. There are plenty of small things – insects, spiders, millipedes, lizards, snakes and many of the smaller birds – but the bigger mammals and birds have gone for ever.

Now, however, they are to be replaced. In the module behind Kule Taaran is the prototype of the new forest creature. Mankind has civilized himself into a synthetic corner: he cannot survive without the full power of engineering science and medical technology. He has turned his back on the natural systems of evolution and ecology that brought him into being in the first place. Now, as the technological systems are beginning to fail more and more frequently, it is time to look back to the natural environments.

The Andlas were overlooked for so long. Once despised

500 YEARS HENCE

FOREST-DWELLER

Homo silvis fabricatus

There is plenty to eat in tropical habitats. The climate is stable and seasons do not regulate the food supply. Like earlier animals that lived there, a human being engineered to live in the abundant rainforest needs only the ability to climb to feed itself. Cunning and intelligence are not necessary – though an instinct for survival is. A level of intelligence will redevelop in *Homo silvis fabricatus* over the coming millions of years, as evolution takes place, but not as much as in species faced with more challenging environments.

Ape-like arms and long fingers allow the forest-dweller to swing in the canopy of the trees; while its strong prehensile toes can grip the branches tightly. A heavy jaw is adapted to cracking nuts.

Although intelligence has been suppressed in the engineering, natural curiosity still comes to the surface.

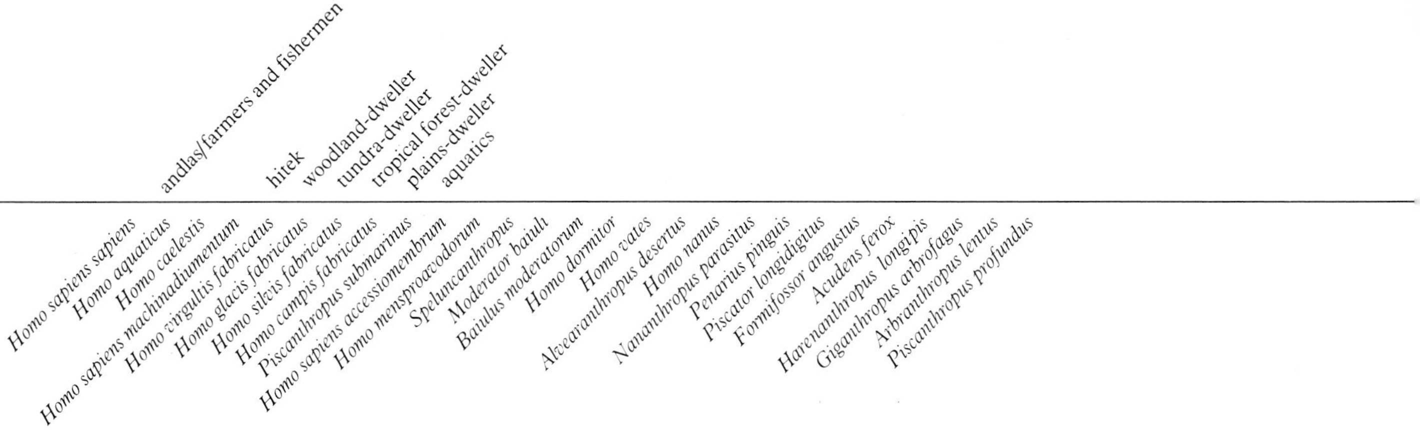

andlas/farmers and fishermen
hitek
woodland-dweller
tundra-dweller
tropical forest-dweller
plains-dweller
aquatics

Homo sapiens sapiens
Homo aquaticus
Homo caelestis
Homo sapiens machinadiumentum
Homo virgultis fabricatus
Homo glacis fabricatus
Homo silvis fabricatus
Homo campis fabricatus
Piscanthropus submarinus
Homo sapiens accessiomembrum
Homo mensproavodorum
Speluncanthropus
Moderator baiuli
Baiulus moderatorum
Homo dormitor
Homo vates
Alvearanthropus desertus
Homo nanus
Nananthropus parasitus
Penarius pinguis
Piscator longidigitus
Formifossor angustus
Acudens ferox
Harenanthropus longipis
Giganthropus arbrofagus
Arbranthropus lentus
Piscanthropus profundus

because of their unsophistication, but tolerated because of their versatility and ability to keep the machines working, they are now recognized as the gene pool for the future of mankind. It is so obvious. Mankind was in a shambles because it had turned its back on the natural process of evolution. The Andlas, however, after the great schism of humanity brought about by the overpopulation and famine disasters of a few centuries ago, had fallen off the hurtling escalator of technological complexity. They represented the part of humanity that had been thrown to the wild and denied the advantages of the constantly-improving technology and culture. These unfortunate creatures lived as best they could, and suffered terribly from diseases and accidents. It was these that reinstated the process of natural selection, and as an ironic result the surviving members became fitter and healthier, generation by generation. It eventually became obvious to the mainstream of technological man, however much his soul rebelled against it, however much his ego denied it, that here lay the purest essence of humanity now surviving.

This is the basis for the humanity of the future. From now on man should not use his science to change the environment to suit himself; rather he should use it to change himself to suit his environment. By his own technological application he can catch up with the thousands of years of evolutionary change that he has forfeited. It is now possible to breed and genetically manipulate new creatures that do not need a technological civilization. Out there, in the tropical jungle, the grasslands, the deciduous woodlands, the coniferous forests and the tundra, are supplies of food growing wild. All the necessities for life are there. If the human body can be regarded as a machine, like a life-support cradle, then the carbohydrates produced in the leaves and tubers can be used as the body's fuel. The proteins in the growing shoots and in the insects can be used as building materials. The vitamins in all living things can be used for lubrication, and the water that is found everywhere can be used for cooling and cleaning. All this goodness was once harvested by a vast range of big animals. There are none left now, and all the food is there for the taking.

Kule Taaran looks at the creature in the transportation module. Strange that this should represent mankind of the future – it looks so much like mankind of the past. The prehensile feet are there, with the big toe modified as a thumb, for climbing and grasping branches. The long ape-like arms with the long fingers will also help it to move about in the treetop canopy. The head seems to be very heavy about the jaw, to accommodate the huge nut-cracking teeth.

The genetically-engineered beings that have been developed for the other vacant habitats seem to be working well, according to the reports. Now they must see how the tropical forest version performs.

The naked form of Pann, sitting amongst the bars and perches of his module, seems ready for his great adventure. He exchanges a few words with Kule Taaran, who then opens the access of the module. Pann leaps from the vessel and into the swaying wispy branches of the nearest bush. He hangs there for a moment while he looks around at the infinite vistas of his new home. Then, with a final wave to those who had nurtured and developed him, he jumps to the nearest tree, shins up the trunk and is lost to sight amongst the branches.

Kule Taaran turns away from the window and back to his console. Physically the new creatures seem to work well; the next stage is to see if they breed true.

KNUT THE ENGINEERED TUNDRA-DWELLER

Mosses, lichens, heathers, coarse grasses – very meagre fare. Yet such a diet used to support very large animals like reindeer, musk ox and mammoth. So there is no reason to suppose that a suitably-engineered human could not subsist on such a diet.

Knut has been raised on it for a decade, but that was in the safety of a living module. The cold-weather plants were brought in regularly by flying machine, and the chill conditions were maintained artificially. All this time, Family have been outside in the warm and looking in.

Now the situation is reversed. The Family members, in their cradles, with all their delicate life-supports, are keeping warm in the modules of the flying machine, while Knut it outside, standing in the crisply-frosted grass of the tundra wilderness, beneath the vast cold grey and white sky. This is what he was brought up for, to take his place in nature.

Centuries ago there were herds of big animals here, which moved north and south as the seasons changed, wintering in the deep forests to the south and spending the summer on these wild plains. In those days, he was told, there were fierce hunting animals as well – animals that would harry and kill the gentle plant-eaters. Now there are none of those left either, and the whole landscape is his.

He looks down at the coarse little plants at his feet; they look the same as those he has been eating all his life. With the ice-hook developed from the nail on his big toe, he scrapes up a patch of moss, then he goes down onto his furry knees and scoops it up with his spade-like hand. Yes, it tastes just the same. He will survive here.

The whiteness that has been building up at one side of the sky descends. Chill flakes of snow begin to swirl past him, settling on his fine-curled fur over the layers of fat. In reaction he rolls up the ruff of fat around his neck and his face disappears into it. From the direction of the great flying ship behind him he can hear the clang and hiss of hatches and accesses closing and sealing. This is too much for Family. There is a sudden blast of warm air as the great vessel leaves the ground. They are going back to the cities where it is warm. Knut is left here, where he belongs.

Yet too much harsh weather will kill him, and the brief northern summer is over. He knows what he must do. As the sudden flurry of snow passes, he brings his face out from the folds of fat and turns it towards the south where away yonder are the huge coniferous forests and winter shelter. Like the great herds of grazing animals before him, he moves southwards with the season.

Yet, unlike the animals of the vast seething herds, he is alone – the only one of his kind – but this does not worry him. If he survives, and he has every intention of doing so, then the experiment will be a success. Others like him will be produced and together they will repopulate the chill northern wastes of the planet.

RELIA HOOLANN AND CULTURED CRADLES

This is *not* the way. Taking Andlas and changing them into wild animals is not the way. Mankind's fate does not lie with these low creatures, but with those who have sustained the technological advances over the centuries. If mankind's future is not one of technological progress, then what is it?

Relia Hoolann and her team have worked for decades on the problem, learning from the genetic experiments of centuries, and at last she has the potential for success. For long enough it has been possible to grow synthetic kidneys, livers, lungs and many other organs. It was the connective tissues and the locomotor systems that were elusive.

It is hundreds of years since a child was born that was free of genetic defects, and able to live without a vast technological back-up (apart from the Andlas, that is; but they do not count). A newborn child must be analysed and diagnosed immediately in order to find out what it needs and to manufacture a cradle for it that contains the mechanical or synthetic equivalents of those organs that are defective – a very clumsy process.

Now it may well be possible for the mechanical parts of the cradle to be dispensed with altogether, so that the whole cradle is grown as a biological unit. What is more, these cultured cradles may well be able to breed, and to reproduce themselves.

This does not spell the end of mechanical technology, however. The process will be very energy-consuming, and the solar-power plants and the ocean-current energy units will be as important as they ever were, not to mention the food carbohydrate and protein factories that will still be required.

This is going to be the saving of the human species, thinks Relia Hoolann. The cultured cradles will be much safer and much more reliable. The fall in population that has been documented over the last few centuries will at last be reversed, now that there is a reliable technology to sustain it.

The first cultured cradle will have to be huge, as there is no way of generating synthetic organs that are as compact and neat as the real thing. In nature most organs have more than one function: you can use your mouth for eating, breathing or talking and your fingers for feeling or handling. The synthetic organs that have been developed can only perform one operation at a time. After all, evolution took 3500 million years to produce natural organs, but humans have only been dabbling with the process for a few hundred.

FIFFE FLORIA AND THE HITEK

Fiffe Floria looks up contemptuously at the ugly unnatural form of the flying ship as it moves silently, blindly overhead and disappears beyond the tall trees to the east. She cannot regard the Hitek, the beings inside it, as human. How can you be human if your life is sustained by mechanical contrivances, and you have to eat food that is made by a machine?

With a dismissive sneer she pulls the coarse veil over her face and tucks it into the fibre belt of her tunic. Then she removes the lid from her beehive and waits for her swarm

500 YEARS HENCE

FIFFE FLORIA

Homo sapiens sapiens

In isolated communities across the Earth, groups of humans have consciously returned to the old land-based ways of living – farming, fishing and gathering. Descended from those who survived centuries of poverty and savagery while squatting in the city ruins, and now abandoned by those who can use their technology to escape, the farmers have proved particularly healthy and adaptable. Now that the population of Earth has fallen to a low and realistic level, the survivors can husband the limited food resources of the planet at a sustainable rate.

Subsistence farming can be harsh and demanding but combined with simple gathering and fishing, it enables small autonomous groups to live in precarious balance with nature.

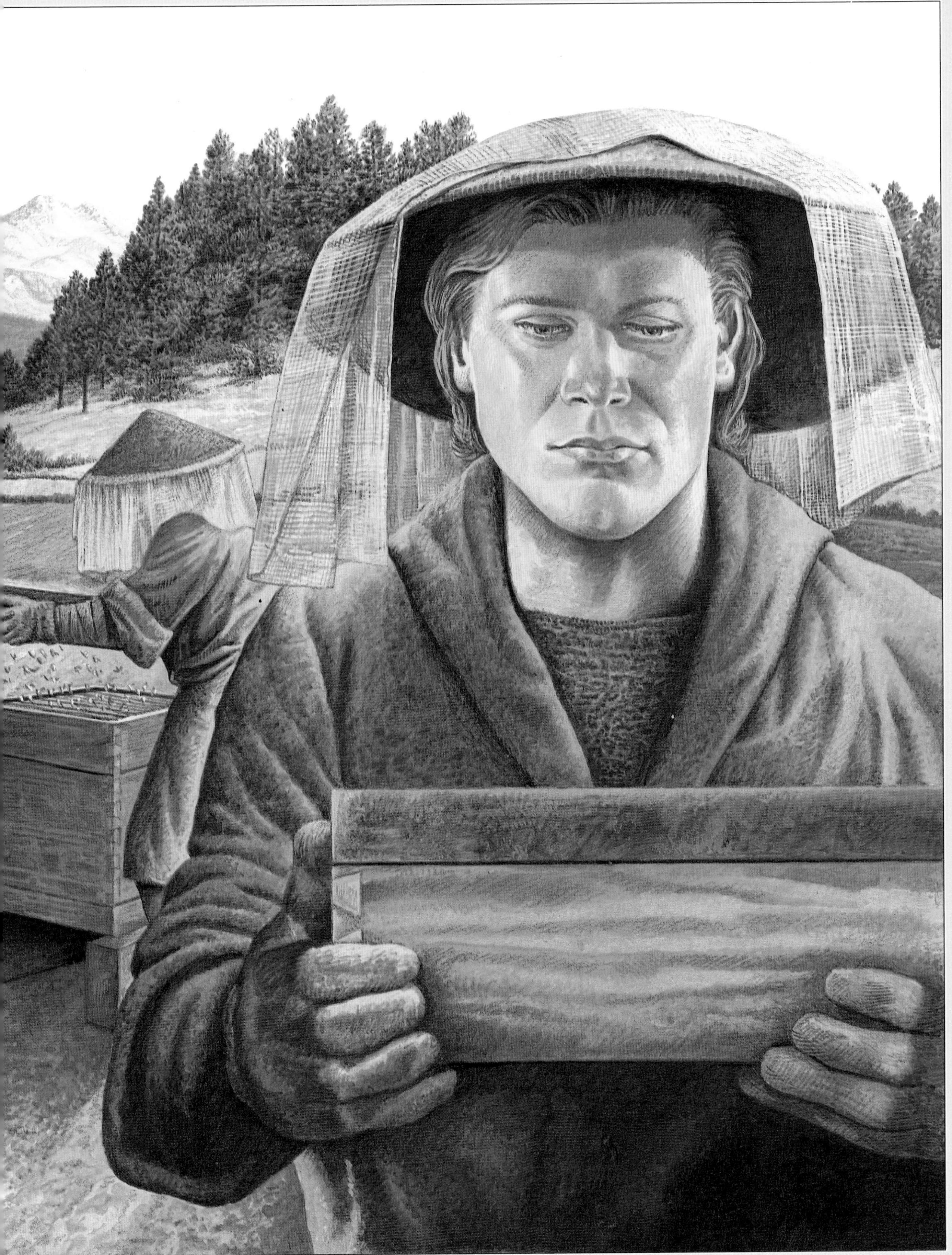

to settle in the smoke from her torch before inspecting the combs. Good. They are filling up nicely, and it will soon be time for the harvest. There seems to be nothing untoward in the hive: no thieving by wasps, no break-ins by mice or rats, no sign that the queen is going to decamp and take half the workforce with her – but it is really past the season for that now. Yes, the harvest is going to be good this year.

Fiffe closes up the hive once more and turns back down the slope towards the settlement. They have been lucky this season. The patch of growing crops is beautifully healthy and the smoke house is full of fish caught in the stream earlier in the summer. Further down the slope lie the overgrown hulks of the great buildings. Once these were completely submerged in the ocean, but now, year after year, the sea retreats further and exposes more of them. This is probably something to do with the climate becoming gentler and cooler. Centuries ago when the world was teeming with people this used to be a great city. It must have been a terrible time, with everybody living on top of everybody else, and no room to expand and breathe.

It may still be like that in the cities of the Hitek. The people in the old cities suffered from lack of food and land, what there was being poisoned. Then the air got too warm, the sea rose and the cities drowned. Deny nature and that is what happens, and it will happen to the Hitek as well.

Her man, Hamstrom, is playing with little Harla on the beaten earth outside their hut, and the beautiful smell of cooking fish is wafting out of the curtained doorway. Harla is their fourth child, and the only one to have lived. They know that she will survive and thrive. The settlement consists of about 100 people, which is just enough for their crop land and their fishing stream to support. If they used the ancient form of measurement they would say that they occupied 50 square kilometres, or a region that was a little less than 5 miles square. Over the hill to the north there is a similar settlement, and to the south another.

To think that the Hitek believe them to be inferior, just because they have not become so inbred and decadent that they need mechanical devices to keep them alive. You cannot live in a natural world by turning your back on nature, regarding it as an inconvenience to be overcome, a hazard to be avoided, an irritation to be shielded against. If that is what they wanted, they should all have gone to the stars on the colony ships of centuries ago. The time will come when they will see that the future does not belong to them, with their artificial systems, but to those who can live in balance with nature.

500 YEARS HENCE

TEMPERATE WOODLAND-DWELLER

Homo virgultis fabricatus

A human-based creature engineered to survive and flourish in a temperate forest without the back-up of civilization would need to be omnivorous. Forests are less abundant than jungle. To reach the full range of foodstuffs available, *homo virgultis fabricatus* has to be extremely nimble, and be able to live both at ground level in the undergrowth and high in the tree-tops. Arms and legs are of similar length and long, but agile, climbing fingers increase its range. A covering of fine hair keeps the woodland-dweller warm in the temperate conditions.

The omnivorous diet is reflected in the dentition, with heavy crushing back teeth for nuts, and delicate front teeth for catching insects. Its diet is close to that of early man; as is its evolutionary potential.

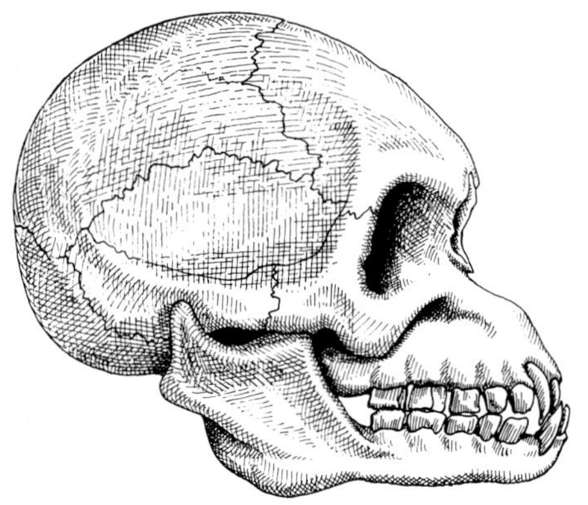

Long prehensile toes and fingers can grip rough bark. Lack of a supporting big toe means that the forest-dweller walks crouched but climbs with ease. It is the least specialized, and therefore, most adaptable of the engineered species.

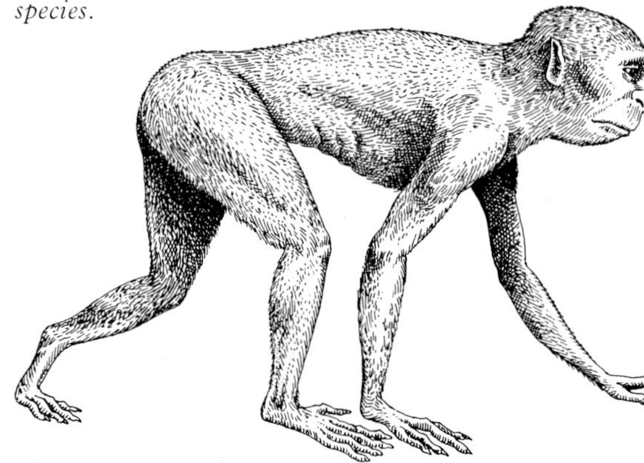

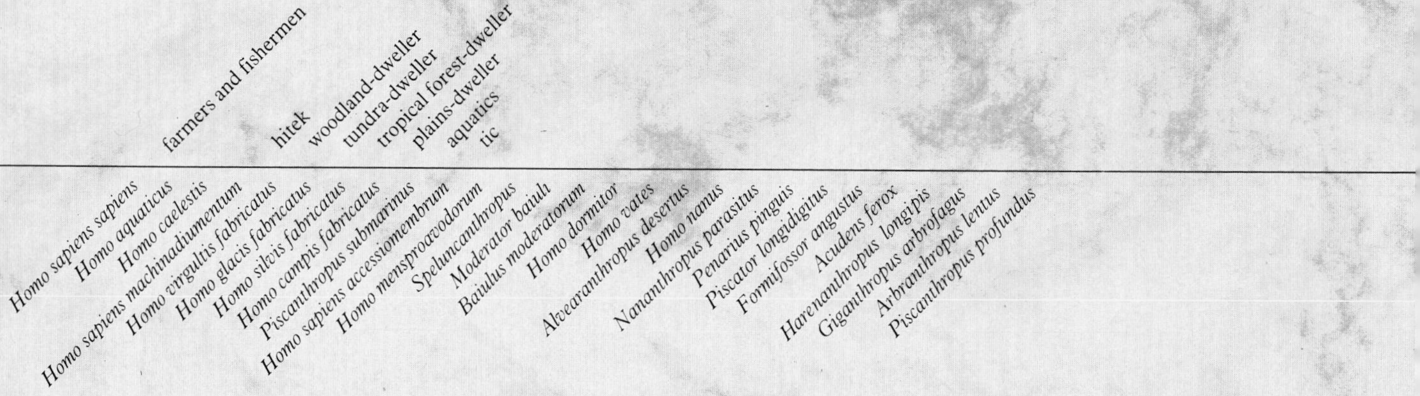

farmers and fishermen
hitek
woodland-dweller
tundra-dweller
tropical forest-dweller
plains-dweller
aquatics
tic

Homo sapiens sapiens
Homo aquaticus
Homo caelestis
Homo sapiens machinadumentum
Homo virgultis fabricatus
Homo glacis fabricatus
Homo silvis fabricatus
Homo campis fabricatus
Piscanthropus submarinus
Homo sapiens accessiomembrum
Homo mensproaevodorum
Speluncanthropus
Moderator bauli
Baulus moderatorum
Homo dormitor
Homo vates
Alveoranthropus desertus
Homo nanus
Nananthropus parasitus
Penarius pinguis
Piscator longidigitus
Formifossor angustus
Acudens ferox
Harenanthropus longipis
Giganthropus arbrofagus
Arbranthropus lentus
Piscanthropus profundus

CARAHUDRU AND THE WOODLAND-DWELLER

This is the one that is going to cause the problems, Carahudru sees that. It has few adaptations, but looks little like its ancestral Andla. It is covered by a fine furry pelt, so that it need never manufacture clothing. The arms are longer, the fingers are more delicate, and the teeth are stronger. What is more the feet are prehensile, with the big toe developed into a thumb to help the creature to climb trees. Deprived of the support of the big toe, it can no longer stand upright, and its position at rest is a four-footed crouch. It is like an animal, but there is no avoiding that. The traditional human frame is totally unsuited for anything but a cultured civilization, so if mankind must live from the fruits of nature without resorting to culture and civilization it is going to have to abandon any traditional view of beauty and elegance. It is going to have to return to the beast.

It is inside the head, however, in the brain, where the most fundamental difference of this creature lies.

What makes a human being? That is the solid residue to which the argument condenses. Are we more human than the Andlas because we make greater use of technology, and they just live in the wild and grow their own food? Are the Andlas more human because they are healthier and look more like our ancestors? If the latter is the case then we could argue that the more primitive the being is then the more human it is.

In that event the specimen before him must be the most human of all. Long arms and prehensile feet will allow it to live both in the deciduous forest trees and on the ground. There is a lack of specialization in the shape, simply because there are so many different food supplies in a deciduous forest that it would not be practical to adapt this creature to exploit any one in particular. However, all this food is not necessarily palatable. Many plant-produced substances are poisonous to the human metabolism, and diseases may abound that have not been anticipated. The engineers have done their best and built in systems that could combat most of the known natural poisons, so whatever has been overlooked will have to be regarded as part of natural selection. As a result, they have engineered a generalized hunter-grazer-browser-insectivorous scavenger.

They also engineered it with low intelligence. If it is to be surrounded by food, the argument went, then it will only need enough intelligence to allow it to find it. An intelligent creature may cause trouble, may feel resentment at being experimented upon, envy at not being able to live in the cities, rebellion against those that engineered it. What is more it may try to better itself, and build its own civilization – and civilization does not now seem to mean a long-lasting and successful species.

In the back of Carahudru's mind is a lurking misgiving. Throughout evolutionary time, the unspecialized creatures proved to be the most adaptable. The new world that is being engineered now is supposed to be balanced, with an engineered creature installed in each environment. If one in particular evolves to encroach on another's environment, what will the long-term result be? It may even be that intelligence will re-evolve by itself.

That is for the future, though. Carahudru throws open the door and his creature steps gingerly out into the bracken and brambles of the deciduous woodland. Immediately it feels at home. Into the thicket it runs, having totally forgotton Carahudru in the flying vessel. Carahudru catches a last glimpse of the sunlight casting a dappled pattern on its back before it disappears into the warm greenery.

1000 YEARS HENCE

KLIMASEN AND THE BEGINNING OF CHANGE

Something is wrong. The ship is not responding properly. Klimasen directs his brainwaves through the neural contacts but they are not having the right effect. The ship is drifting out of control.

He noticed it on the last trip, but not as strongly as this. Then he was able to bring the vast vessel into dock safely, and deliver the food with no real problem.

As a desperate measure he disconnects the neural system, and with his most delicate pair of synthetic arms he removes the guard panel of the instruments before him. He can detect nothing wrong, nothing malfunctioning. Yet still he is drifting away from his course, out of control. If there is nothing wrong with the ship, it must be something external.

Beneath him he can see brilliant white flecks on the grey of the ocean surface – icebergs. He has never seen them so far south before, but that is not really surprising, for the ice has been pushing southwards further each year as the weather has been cooling. That should not worry anybody because the whole of civilization is well guarded against changes in the climate. Only those scattered tribes of primitives will have problems.

The presence of the icebergs does not disturb Klimasen. What *is* alarming is the direction in which the vessel is travelling over them. It is evidently the guidance system that is causing the trouble; but that cannot be – the guidance is worked from the Earth's magnetic field. A surge of alarm sweeps through Klimasen's puny body, and is instantly neutralized by a burst of sedative generated in the bulbous stack of synthetic glands grafted to his back. If the Earth's magnetic field is varying beyond the limits that the machinery of the ship can tolerate, then there may be trouble for all trade and communication around the world.

It cannot possibly be that, he thinks. More likely it is the strong winds that are accompanying the edge of the ice pack; but the sensors do not show any winds that are stronger than expected. Something is seriously wrong.

Desperately he brings his manipulator hands into play to work the seldom-used manual controls, but that does not have any effect either. The ship is descending at a great speed, faster than he can correct it. Even if he could stabilize it, there is no way of telling which is the right way to go. He is totally lost.

Cold grey ocean and glistening icebergs are rushing up to meet him. With his feet and his most powerful pair of arms he braces himself for the impact.

THE END OF YAMO

For the tenth day in succession the clouds have obscured the mountain top. The sunlight that does filter through is not enough to activate the solar collectors and keep the food generators working at full efficiency.

For the first time in his life Yamo finds his work overwhelming, and his efforts largely fruitless. *He* does not control the process. He just inspects the machinery that repairs the devices that do control the process. He does not think that there is anybody now living who knows enough to control the process, and now this particular plant is collapsing because the machinery is slowing to a halt. There is no power coming in from the solar collectors, nor is there any coming in through the network from other collectors in other areas. Everybody else is having the same problem. What is more, power-storage facilities are almost exhausted.

His massive carrying legs transport him, cocooned in his organic cradle, down to the depths of the factory. He has lost count of how many times he has made that journey in the past few days. It is all to no avail, as there is nothing he can do when he gets there. It is still as silent as ever, but the smell of decay, as the nutrients and raw materials rot, is stronger.

There is something disrupting the weather systems, something that was never allowed for when the manufacturing process was designed. All right, the climates are gradually becoming cooler as time goes on, but this is a gradual process, and something which was taken into account when the whole system was set up. It should not bring about the effects that are being produced now.

His food cake appears at his dispenser. At least, working in the plant, he has first call on what food there is left.

The door hisses open. Someone else stands there, someone he does not recognize. The light is behind the figure and all that Yamo can make out is the silhouette – the lumpy shape of a standard organic cradle, with the powerful legs, and a selection of arms dangling.

What is this person doing here? No-one has ever come into his module before. It must be important. Then he realizes that with the power deteriorating the communications systems must be failing as well. There has been no communication from outside at all for days. He turns to check his screens and monitors, but before he can do so he feels a pair of handling arms seize him. Manipulators rip into his own cradle, reaching for his head.

Dimly, as Yamo's biological back-ups rupture and collapse in a spray of blood surrogate and synthetic hormones, he realizes that he must be the first murder victim for centuries.

Murder, too, for the oldest of causes. The newcomer steps over the pulsating form of Yamo's broken cradle, and picks up the little cake of food.

WEATHER PATTERNS AND THE TICS

They laughed when it first started, the farmers and fishermen. They could see that the ocean currents were changing. They knew that somewhere out there, at a great depth below the sea surface, was one of the great ocean-current power-generators that supplied the energy for the Tics. Now the movement of the water had changed and it would

THE TIC

Homo sapiens accessiomembrum

Medical technology has developed 'soft' forms of the back-ups that keep alive the weakening human form. Replacement organs, grown synthetically, are grafted onto the body. Eyes, ears, mouth and nose still function. The fingers work only as organs of touch. Lifting or handling is left to arms grown artificially. Fashion plays a part in such surgery.

Genetic engineering is not so far advanced that something grown artificially can match the complexity of 3500 million years of evolution. Grafted organs are single not multifunctional.

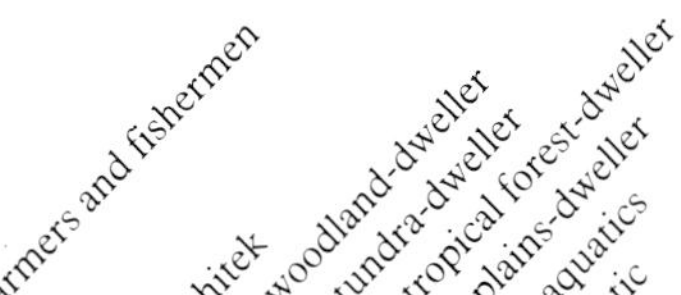

not be working any more. How were the Tics going to keep themselves alive now, in their monstrous living suits and their food factories?

Now, however, it is not so funny. The new weather patterns have brought unceasing rain, and the crops have failed. The fish have not come to the river this year, as though they could not find their way to the spawning ground. The bees are in disarray; they cannot see the sun and their internal directional instinct is failing them.

It seems to be happening throughout nature. Every year the birds move north and south at the same time, but not so this year; they do not seem to know their directions. It is affecting people, too. The trading caravans that move between settlements are becoming confused and lost. Men and women admit that they are finding it difficult to find their way along even well-known routes.

Then there are the sicknesses. Diseases that have never been known before are beginning to afflict those who spend their time outside. It seems to be something to do with the sun, which whenever it appears from behind the unfamiliar clouds is harsh and glaring. It burns the skin, and produces growths that do not go away until the victim dies.

The collapse is coming all right just as the farmers and the fishermen have always predicted; but it is not restricted to the Tics. It is going to affect everybody: those who deny nature *and* those who live with it.

PLAINS-DWELLERS

The hazy grassland stretches away, green and yellow, to infinity, and the herd of grazing creatures moves gracefully across it. There are about 20 of them, the adults moving along on the outside of the group, with the youngsters in the centre. This is some sort of instinctive arrangement, serving no real purpose, as there are no dangerous animals to defend against. They have no real speech, these creatures, since all their needs are simple and amply met. Food grows all around, there are no enemies, and they have the companionship of their own kind.

Towering clouds are building up overhead. The plains-dwellers are aware, but only dimly, that conditions are changing from year to year. There seems to be more rain than there used to be, but this no problem. It only means that the grass – their food – grows more prolifically. It also means that new types of plants are beginning to grow: saplings that will develop into bushes and trees. Still, there will be plenty of grass left for them.

As they move slowly through the waving leaves and stems, they become aware of a distant humming noise. Looking up, their leader sees an oval spiky shape floating above the horizon away to one side. Such things go over now and again, but they have no effect on the plains-dwellers, who barely notice them.

However, this one is different. It is not pursuing its usual straight unwavering course but seems to be tilting to one side and descending in a very irregular manner. This is unusual enough for the herd's leader to stop and look at it, as does the rest of the herd.

The shape wobbles, and finally drops into the plain some distance away. Immediately it is engulfed in a white flash that fades into a billowing red and black ball, rising and spreading. A little time later, the explosion is heard as the sound sweeps across the open countryside, and the infants and parents alike start in alarm, but feel no fear. The leader, however, does see the danger. The burst of red has spread as a fire across the landscape and it is coming towards them.

He has seen this before (fires are commonplace on the grasslands), and is knowledgeable enough not to run away from it when it is sweeping towards the herd. He assesses the direction of the wind and moves his herd along at right-angles to it, so that the fire will eventually travel by them.

He need not have troubled. The clouds that have been building up throughout the afternoon now open, and a curtain of torrential rain appears between the herd and the fire drifting over them and soaking them instantly. By the time the downpour has passed nothing is left of the fire but a steaming black smudge on the distant landscape.

The erstwhile flying shape is steaming and black as well, but the plains-dwellers ignore it and continue their journey. It has nothing to do with them.

HOOT, THE TEMPERATE WOODLAND-DWELLER

The advantage of living in a temperate deciduous forest is that there are so many different things to eat at different times of the year. In the spring there are delicate shoots and soft buds; in the summer, the trees and bushes are full of leaves; and autumn is the time of fruits. It is winter that gives the problems. With any luck a forest-dweller has eaten so much throughout the rest of the year that it has built up enough fat to enable it to exist through the lean months, or it may be sensible enough to gather food such as nuts during the autumn and store them away for winter.

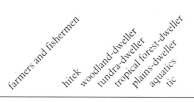

Throughout the year, too, there are insects, grubs and small animals hiding under stones and beneath the bark of trees.

The temperate forest-dwellers were designed as omnivores, in order to take advantage of all these circumstances.

Hoot is typical. He looks very much like his great-great-great-great – great to the power 20 – grandfather, who was one of the first genetically-viable forest-dwellers to be engineered. He is built as a climbing creature, with long arms and legs, but he is just as comfortable on the ground. His teeth are quite generalized, able to cope with a wide range of foods from soft fruits to hard insects. His main senses consist of sight, smell, taste and hearing.

In fact, in outward appearance he resembles the ancestral human being. Inside his long body, however, his digestive system contains special organs for treating particularly tough food, and self-sustaining colonies of specialized bacteria that can break down tough silica and cellulose, allowing him to digest just about anything that he swallows.

His mind, though, is dull. That was part of the plan as well, as it had been believed that such a creature would survive better without the typical human power of logic and reasoning. Its food was all around it, so it would not need to experiment, to try to make its life more efficient, since its environment would sustain it perfectly adequately. The prototype worked so well that many others were engineered, and now there are self-sustaining colonies throughout the temperate forests of the northern hemisphere.

Nevertheless, Hoot now finds something new in his forest. On top of the hill, close to his own trees, there has always been an array of glistening things, like the leaves of a tree, but bigger and square. Hoot has always known that something big exists deep within the hill, connected to these strange things. A minor sense that came to the surface when his ancestor's furry pelt was engineered was sensitivity to electrical fields: a tingling of his hair roots tells him when he is in the presence of electrical machinery. He understands none of this, of course, but he knows that this sense tells him that something important lies beneath the hill; and this something big is important to the lumpy creatures that he has always thought of as some kind of distant relation to his own people.

An unfamiliar noise and increased electrical disturbance has brought him to the hill this morning. Flying things came in from all round the sky and descended, disgorging more lumpies than there are woodlice in his tree. Sometimes when his own people are angry with one another – say, if he wants to mate with the same female as somebody else – he can sense the tension in the air. Anger and hatred are obvious and can be communicated without noise, and it is the same here. Hundreds of lumpies have collected together and they are angry. They want to get into the hill, and are pushing at the door.

Eventually they break through, and other lumpies come out and tackle them. Hoot has never seen such a fight – dozens of lumpies tearing away at one another, pulling each other apart, stamping each other into the ground. His own people do not do things like that.

Eventually the battle moves on, into the hill. The noise and the chaos retreat underground, leaving the soil littered with dead.

Stealthily, Hoot descends from his tree and scampers over to the site. The first dead lumpy is still warm, and oozing blood. He sniffs all around the corpse, using his selective sense of smell to ignore the main odours and concentrate on the smells that seem most interesting. He lowers his mouth to the seeping wound and, experimentally, licks the blood. It is good. He licks some more, using his tongue as an organ of touch, to find the parts that may be palatable. Then he starts drinking.

His digestive system was designed to absorb almost anything. This is as big a feast as he has seen in many a day, and the others of his kind should have a part of it.

Rearing up to his full height, he lets fly his own recognition yodel, summoning all of his brethren who are within earshot. It looks as if this is going to be an easier winter than last.

As he hears the crashing and scampering of his relatives approaching through the leaves and undergrowth he turns back to his feast. With a feeling of contentment he sinks his teeth into the synthetic flesh and artificial organs of the creature before him.

THE END OF DURIAN SKEEL

Some things just cannot be predicted, Durian Skeel muses; but he knew that the end would come as something like this. Mankind has built defences against everything that nature can inflict. Throughout human history the waste products of civilization built up and poisoned the air, the seas and the land. When the damage became too much to bear, technology was brought in and in the end halted the

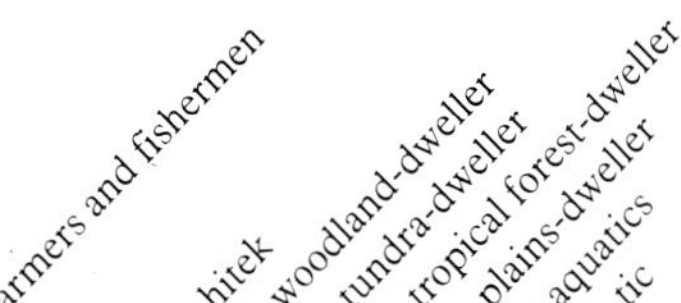

Homo sapiens sapiens
Homo aquaticus
Homo caelestis
Homo sapiens machinadumentum
Homo virgultis fabricatus
Homo glacis fabricatus
Homo silvis fabricatus
Homo campis fabricatus
Piscanthropus submarinus
Homo sapiens accessiomembrum
Homo mensproavodorum
Speluncanthropus
Moderator baiuli
Baiulus moderatorum
Homo dormitor
Homo vates
Alvearanthropus desertus
Homo nanus
Nananthropus parasitus
Penarius pinguis
Piscator longidigitus
Formifossor angustus
Acudens ferox
Harenanthropus longipis
Gigantanthropus arbrofagus
Arbranthropus lentus
Piscanthropus profundus

process. Nature repaired the damage eventually. Now processes have been found that produce little or no waste; but it has not been enough.

Climates have been gradually changing for ages. Now mankind can shelter away in artificial habitats, immune to the changes in weather conditions; but it has not been enough.

Only so much food can be grown or manufactured. The only way to guard against shortages has been to regulate population, so that there are never too many people for the available resources; but it has not been enough.

There are the larger-scale processes that mankind can do nothing about, no matter how sophisticated the technology. The moon goes around the Earth. The Earth goes around the sun. The movement of the Earth's metallic core generates the magnetic field that has subtle influences on everything on its surface.

It has always been known from the geological record that the magnetic field changes. At times in the past there has been a magnetic north pole at the geographic north pole and a magnetic south pole at the geographic south. At other times there has been a magnetic south pole at the geographic north and vice versa. It has never been fully understood how these change, when they change and how long the change takes to occur. There must be times, during the changeover, when there is no magnetic field whatsoever, and this must have an influence on just about everything.

The Earth is undergoing just such a change now, and there is no magnetic field. The most obvious effect is on the technology of transportation and navigation. With no magnetic field the compasses and everything that works on a compass principle must cease to function. There are natural processes of navigation as well: most creatures have organs, sensitive to the Earth's magnetic field, which help them to find their way about. The mechanics of fish and bird migration and the homing processes of bees have been disrupted and are now ceasing to work.

Humans have this ability too, but it has never really been used. Only now that the field has collapsed is its absence noticed, with even the most sensible and level-headed of people becoming confused about direction and time and many other subtle things. In the natural world this should not really matter, since the magnetic effect is relatively minor, and most animals navigate by the sun and the stars. However, with no magnetic field the ozone layer of the atmosphere breaks down – just as in the bad old days of pollution. This allows for deeper penetration into the atmosphere by ultra-violet solar radiation, upsetting the normal climatic patterns and producing abnormal wind circulation and hence abnormal ocean currents. The resulting overcast skies break down any biological stellar navigation systems.

On top of all that, there is the harmful biological effect of ultra-violet rays: burns and skin cancers develop wherever the sun does shine through, and birth abnormalities are increasing to well above normal levels. Then there is the disruption of radio waves through cosmic interference. Each human community is now effectively isolated from any other – denied both the exchange of information and physical travel.

Modern civilization and technology are not tuned into any of this. Durian Skeel knew that all this was going to happen, and he tried to warn people from the start. They would not listen.

He takes a grim satisfaction in the knowledge that he, and only he, foresaw the collapse of human civilization. Its death would be slow, from a human point of view, but rapid and catastrophic in the historical scale. Eventually the magnetic field will re-establish itself, with the opposite polarity to before. It may be within months, or it may take decades, but it will be too late to rescue civilization as it hurtles downwards into rubble.

He is not waiting. Purposefully and methodically he disconnects each of his life-support devices and lapses into peaceful oblivion.

AQUAS

Beneath the tumultuous surface of the ocean, the aquas swim around in a leisurely fashion. Something is different, but they do not quite know what. The huge machine with its constantly-turning rotors and fans is now still and silent for the first time in memory. That is nothing to do with them – it was built by the strange beings from above the surface. The movement of water is different, but that has

The earth's electromagnetic field fails as the magnetic poles reverse. On land migration ceases and at sea, as a result of changes to the ozone layer, the ocean currents change as wind patterns are altered. Beneath the waves, giant generators fall silent to be colonized by sealife.

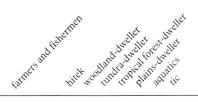

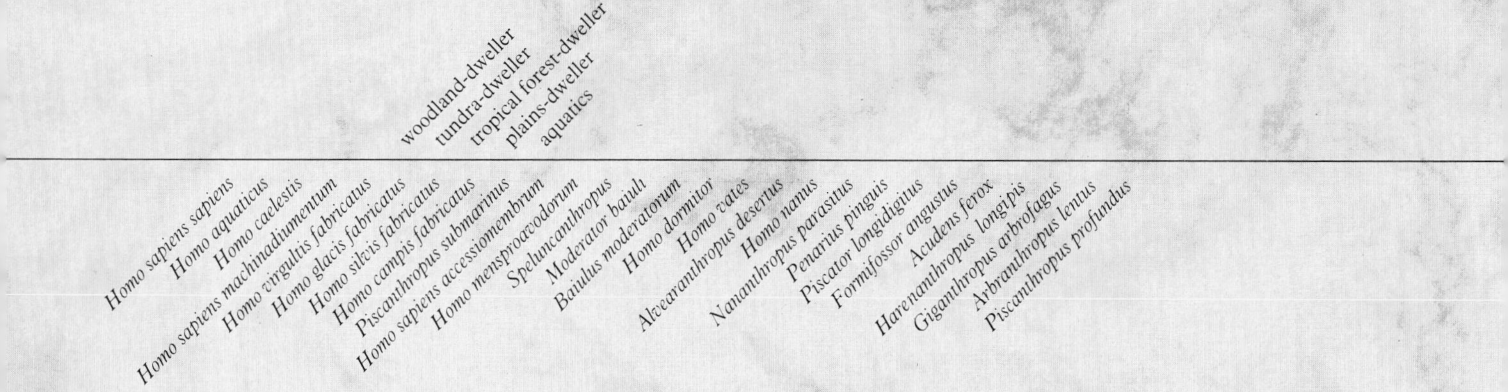

woodland-dweller
tundra-dweller
tropical forest-dweller
plains-dweller
aquatics

Homo sapiens sapiens
Homo aquaticus
Homo caelestis
Homo sapiens machinadiumentum
Homo virgultis fabricatus
Homo glacis fabricatus
Homo silvis fabricatus
Homo campis fabricatus
Piscanthropus submarinus
Homo sapiens accessiomembrum
Homo mensproavodorum
Speluncanthropus
Moderator baiuli
Baiulus moderatorum
Homo dormitor
Homo vates
Alvearanthropus desertus
Homo nanus
Nananthropus parasitus
Penarius pinguis
Piscator longidigitus
Formifossor angustus
Acudens ferox
Harenanthropus longipis
Giganthropus arbrofagus
Arbranthropus lentus
Piscanthropus profundus

no effect on them either. The fish and the sea plants are still there. Even now the sea life is beginning to colonize the vast dead structures.

This may be a good thing for them. They do not now need to travel so far to find their food, and the new children that are born seem to have a better chance of survival now that food is more available. What is more the knowledge has gone out across the seabed, and aquas from other areas are moving in. It looks as if the population is growing quite fast in this area, and they no longer travel in small family groups. A whole interactive society may develop in this region, with all the advantages which that entails. Things may change from now on.

2000 YEARS HENCE

RUMM THE FOREST-DWELLER

Still it becomes colder. This is obvious even to a being of such dim perception as Rumm. His favourite hollow has not yet cleared of snow, and already the sun has passed its high point. From now on, for the rest of the year, the days will become shorter and the air colder. Therefore the snow will not melt at all.

It is going to be far more difficult to find food. Although his intellect is basic, his senses are acute enough when food needs to be located.

His mate and his children are safe from the cold in their hollow-tree den, but they will soon be hungry. They may need to move away, to follow the sun like all the others in the area have done. Rumm has always resisted that because his instincts told him that if everybody else moves away all remaining food will be left for himself and his family. So far this philosophy has worked. The gathering of food has become more and more difficult, but there has been enough to keep them alive. Now he is not so sure. If the snow does not melt, then little will grow during the rest of the year.

He gathers the twigs and branches of the scrubby bushes rising above the snowy ground cover. With a prickly armful he turns back towards his den. The leaves will be bitter and tough, but at least they will be edible.

He surmounts a ridge, and glimpses a group of people below him.

Fast as a blink, he drops his branches and falls to the ground, off the skyline. What are people doing here? Everybody in the area has moved away, following the sun.

Stealthily he moves back up the slope and peers cautiously over the top. They are people, all right, but quite unlike any people he has ever seen. Their bodies are padded out with fat, and their hair is dense and curly. There are thick rolls of flesh around their necks and wrists, and their faces are broad, with enormous nostrils but tiny eyes. There are about ten of them and they are moving towards the sun.

It is as if these creatures have come from an even colder place, and they are following the sunshine, just as Rumm's people have done.

Are they really people? They have a body, two arms, two legs, just like Rumm, but apart from that they are quite different, with their furs and their fatty rolls. They are also from a different place, so they cannot be people, like him.

They must, then, be animals.

Food!

The coarse leaves and twigs forgotten, Rumm waits until the group has crossed the open ground and moved into the wispy trees. Then he scampers down the slope towards their track, taking advantage of any cover that lies in the way. Their smudgy footprints in the snow make their trail easy to follow. Silently, as he does when he is stalking birds, he creeps up on the rear of the party, waiting for stragglers. There are none. They are moving as a tight compact group.

After a while the party comes across the stream that runs through the valley, tinkling along coldly between transparent shelves of ice. They pay it no heed, but move onwards, except for one youngster. Unnoticed by its group, it goes down onto its knees by the chilly water, scoops some up in its broad palm and begins to drink. The remainder of the group presses onwards.

This is Rumm's chance. Silently he pounces upon its back and the youngster stiffens beneath him and gives out a single, high-pitched plaintive yell, like one of his own babies crying.

That yell almost stops the attack, so human is it; but he presses home his advantage. Throwing his hand over the creature's broad nose and mouth, stifling the unnerving noise, he wrenches its head backwards, into the folds of its neck. A cracking noise tells Rumm that the move has been satisfactorily fatal, and the body goes limp.

The yell has alerted the rest of the group, who turn back and with cries of anger descend upon Rumm and his victim. It is too late, though. The forest-dweller has hois-

TUNDRA-DWELLER

Homo glacis fabricatus

Mosses, lichens and heathers provide the slow-moving tundra-dwellers with their diet. A hook-like nail on the foot, developed from the main toe, scrapes up moss and also provides a grip on the snow. Migratory by nature, the dwellers move to open tundra each summer but winter deep in the forests. As with all migrations it is the old, the weak and the young who fall prey to predators.

The five engineered forms do not perceive each other as members of the same species. When different types meet, they do so as competitors and enemies; or else ignore one another as irrelevant.

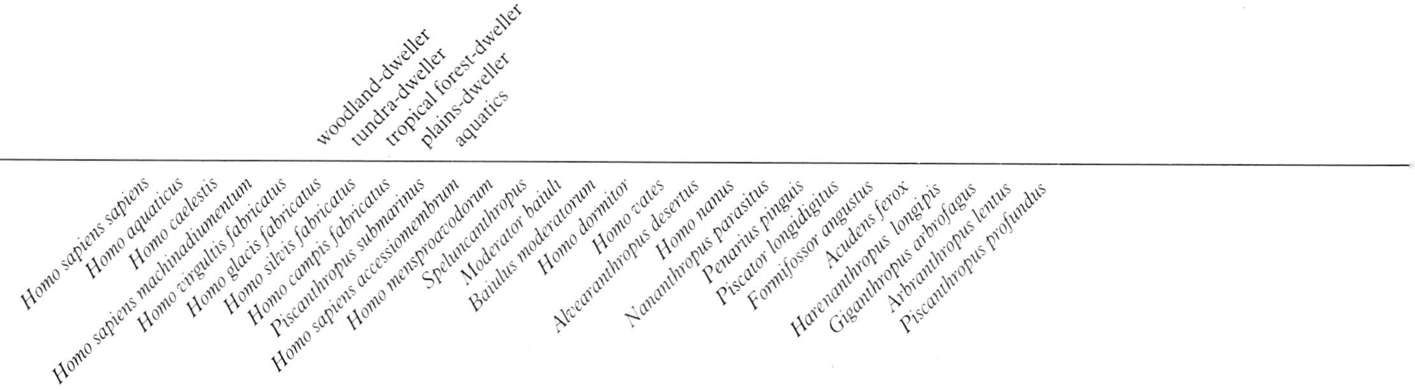

woodland-dweller
tundra-dweller
tropical forest-dweller
plains-dweller
aquatics

Homo sapiens sapiens
Homo aquaticus
Homo caelestis
Homo sapiens machinadiumentum
Homo virgultis fabricatus
Homo glacis fabricatus
Homo silvis fabricatus
Homo campis fabricatus
Piscanthropus submarinus
Homo sapiens accessiomembrum
Homo mensproavodorum
Speluncanthropus
Moderator baiuli
Baiulus moderatorum
Homo dormitor
Homo vates
Alvearanthropus desertus
Homo nanus
Nananthropus parasitus
Penarius pinguis
Piscator longidigitus
Formifossor angustus
Acudens ferox
Harenanthropus longipis
Giganthropus arbrofagus
Arbranthropus lentus
Piscanthropus profundus

ted the dead creature onto his shoulder and disappeared into the snowy thickets. As he goes, he hears the noises of anger behind him, and hears them change into wails of anguish and loss.

What has he done? Creatures that can feel loss so acutely, and can make such sounds of despair – surely they must be people after all? The wailing fades and disappears behind him, but it will remain long in his memory. It will come back to him in quiet moments, or when he is concentrating on something else; and for many a day he will feel sorrow and sympathy with these strange beings. *What has he done?*

He has fed his family, that is what he has done. With a more confident stride he makes his way with his prize back to his mate and his children in their hollow-tree den. They will see the winter through all right now.

LARN THE PLAINS-DWELLER

Larn strides across the grassy plains at the head of his tribe. Not far off he sees a thicket of bushes and thorn trees that he does not trust. Another group of plains-dwellers met danger at such a clump not long ago when a pack of some new kind of animal burst from within, taking them by surprise, and killing three of their number before the rest could escape.

Larn had thought about this incident for some time, and it made him uneasy. He had noticed that the other animals, the little animals of the grassy plains, had their enemies. There was always strife and death in the undergrowth, but not for the plains-dwellers. He had always assumed that this was because the plains-dwellers were the largest creatures around. They had no enemies. The plains were theirs, and theirs alone.

As a result the populations of plains-dwellers are growing and growing. As a lad, Larn could remember travelling with his tribe for days on end, and not meeting any others. Now other tribes are seen daily, and each one seems to be becoming bigger and bigger.

In one part of his mind Larn feels pride at this; his people are the masters of this landscape, and they should spread and fill it. Another, quieter part of him rebels, however. If there are more and more plains-dwellers as time goes by, will there always be enough grass to feed them all?

He turns and looks back at this tribe, and counts them: ten females, all his mates; five young males, that have latched on from other tribes; six of his children, almost adult; twelve of his juvenile children; and two old females, members of the original tribes of two of his mates. He took responsibility for these when he chose the females from those tribes.

It was the two old females that kept the tribe moving slowly. They all had the long legs with muscular thighs and tapering feet that allowed them to run quickly. However, they rarely had the chance to do so. Sure enough the youngsters would run about, very actively, but the older members had to remain close to one another, and so moved at a slow and sedate pace. It was so long since Larn had run that he thought he might have forgotten how – not that there was any real need for speed.

The children enjoy it, though, he muses as he watches them scamper and gambol through the long yellow grasses of the open plain.

Suddenly there is a hideous howling and baying noise from the suspicious thicket. He had let his mind wander and had forgotten the danger that the other tribe had faced.

With a yelled warning he brings the whole tribe together, but the youngsters are scattered too far. A crashing noise issues from the thicket and about ten indistinct forms burst out and streak through the grass. One of his children is brought down with a crash and a flurry of dust and broken plant stems.

Without thinking, for the moment, of his loss and grief he runs about, rounding up the others, trying to get them to crowd together, instinctively knowing that a large group is stronger.

He is dimly aware that the others are doing their best as well. The young males have rushed together in defence of the younger females and the juveniles. They stand shoulder to shoulder while the others sprint into the distance.

Then he comes across a horrible sight. One of the old females lies dead, her throat torn. Over her stands a hideous and misshapen, yet strangely familiar, figure. It is almost like a plains-dweller, but it does not have the long legs, its belly is not so round and its teeth are not so massive. These must be the strange new creatures that have moved onto the plains.

It is staring at him, the female's blood dribbling down its chin. Its eyes are grey and steady, it bares its teeth, and then it pounces. As a reflex, Larn brings down the cutting edge of his left hand, thrusting it into the soft flesh of the creature's neck, killing it instantly. So they are not invulnerable, Larn thinks with triumph; we can kill them.

Then another dark shape crashes into his back, sinking its teeth into his neck, and as he falls into the dust he real-

AQUATIC

Piscanthropus submarinus

Developed in the earliest centuries of genetic engineering as a refinement to the aquamorphs, the aquatics were the first group to carry hereditary genetic changes. Clumsy and vulnerable on land, the sea is now their instinctive habitat. *Piscanthropus submarinus* can move swiftly and powerfully within water. The ocean provides food and does not vary its temperature as swiftly as air – valuable when the increasing cold forces land-based species such as *Homo virgultis fabricatus* into adaptation or retreat.

Even with long toes and fine balance, the temperate woodland-dweller has to move carefully across the slippery rocks. Curiosity proves stronger than its fear of falling.

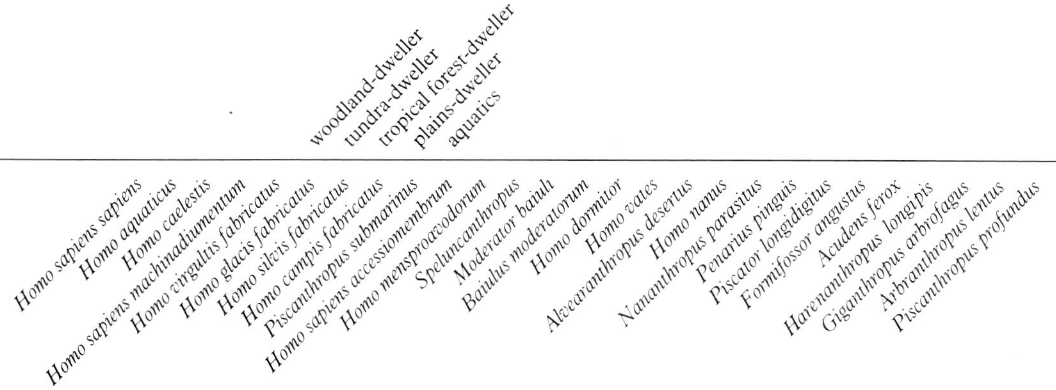

woodland-dweller
tundra-dweller
tropical forest-dweller
plains-dweller
aquatics

Homo sapiens sapiens
Homo aquaticus
Homo caelestis
Homo sapiens machinadiumentum
Homo virgultis fabricatus
Homo glacis fabricatus
Homo silvis fabricatus
Homo campis fabricatus
Piscanthropus submarinus
Homo sapiens accessiomembrum
Homo mensproavodorum
Speluncanthropus
Moderator baiuli
Baiulus moderatorum
Homo dormitor
Homo vates
Alvearanthropus desertus
Homo nanus
Nananthropus parasitus
Penarius pinguis
Piscator longidigitus
Formifossor angustus
Acudens ferox
Harenanthropus longipis
Giganthropus arbrofagus
Arbranthropus lentus
Piscanthropus profundus

izes his mistake. He should have run, like the young females. These creatures have cunning and hunting skill, but they do not have speed.

If plains-dwellers are to continue to be the masters of the plains, they must learn to keep clear of these monsters. Speed is going to be their saving, but it is too late for him.

COOM'S NEW FRIEND

The tide seems to be going out further these days. Coom is only a young lad, but he is sure that he can remember when the water came right up to the cliffs. Yes, sure enough, there is still a line of whitened tree-trunks and bleached sticks, the remains of debris brought up by the waves long ago. His father is much older than he is, and can probably remember when the sea came right up to the foot of the cliff all the time. He might even remember it washing to the top of those austere stone faces.

Now the water is well out, leaving pools and puddles amongst the slippery, weed-covered rock. It will return, before the day is out, but it will not come anywhere near the cliffs. Coom thinks that it probably never will again.

He drops to all fours by the nearest of the rock-pools. The empty woven-reed bag flops onto the cold rocks beside him. Nothing much in the water here. Further down, towards the edge of the sea, the pools will be more alive.

Here he has to be careful. The rocks are wet, weed-coated and slippery; and they are very cold beneath his feet. Now the cracks in the rocks are full of winkles, limpets lie flat and immobile against the wet algae-clad stone, and crabs scuttle and hide in the clear waters of the pools. With his long fingers, Coom pulls the shelly creatures away from their rocks, and dips into the cold waters for the crabs and sea-anemones. It is meagre fare, and even when his bag is full it will not give very much nourishment to this family.

He straightens up and looks back towards the cliff. There, in one of the caves along the foot, live his parents and his three brothers and sisters. It is a good thing, he thinks, that the sea does not come up to the cliffs any more. He and his family would be washed away.

He is far enough down the beach now to see the mountains rising beyond the cliff. They are white, and have been for some time. He can remember, when he was very very little, that sometimes they were green and purple. It is snow and ice that covers them, he knows that. Even the rocks and the cliff are covered in snow and ice now and again. Then a sudden thought strikes him – snow and ice are made of water, so could it be that, with so much more snow and ice over the land, the water has been taken from the sea – and that is why the sea does not come up to the cliff any more?

A loud splash from behind him breaks his train of thought. Something big trapped in a pool! He turns quickly. At first he thinks it is a fish, but he has never seen a fish as big as that. Then he thinks it is one of his family who having slipped in is finding it difficult to get out, but no. It is neither of these.

It seems to be something in between.

The creature rises half-way out of the water. It has a face like his, with eyes, a nose and a mouth; but the eyes are enormous, the nose a pair of slits, and the mouth a vast downturned feature between huge fleshy lips. It has arms and hands, but the rest of the body is indistinct in the water. It seems to be smooth and shining.

Coom stares at the apparition, and it stares back at him. The great mouth begins to work, and sounds come out. It is trying to say something.

Is it dangerous? No, Coom does not think so; in a strange way it is almost like himself. He says a few words back to it, one or two of the few words that he and his family use, but that is no good. Whatever it is does not understand. Instead Coom tentatively reaches out his hand; the odd creature reaches out its own hand, and the two touch.

A friend! Coom has found a friend outside his family.

He lets drop the strange slippery hand, and turns to run back to the cave to tell everybody, full of joy and surprise at his discovery. His father is there, at the entrance, cracking open and scooping out a shellfish that the others of the family have brought him. Coom goes running up to him, grunting out his news. His father is all attention, as are his older brothers.

The result is unexpected. Coom is snarled at to move out of the way, then thrust into the cave while the others run off down the beach towards the sea.

That is not right, thinks Coom, that is not how it should have happened. They do not seem at all pleased about the new friend. He is not going to stay in the cave while all this is happening, so he runs down the rocks after them; but he is too late.

Already his father and his brothers are throwing rocks and bleached sticks at his new friend, and shouting the most hideous threats.

The strange creature, in panic, has pulled itself out of its rock-pool, and is wriggling its way across the clammy weed and cold rocks towards the waves in blind terror, bleating out strange sounds as it goes. Coom stops. He does not want to be any closer, and see in more detail. He can imagine the weals and bruises on the glossy body, the blood from the fresh cuts, the look of anguish and pain on the outlandish face. He can only hope that the strange being reaches the water before his father and his brothers.

With sadness he watches it slip into the waves, beyond the gesturing figures of his family. A flip of the fin-like tail and it is gone.

Well, his father must be always right. Coom considers the matter. He must have done wrong to try to befriend it in the first place. It is obvious that his people, the people of the land and the creatures of the sea will never be anything but enemies.

YEROK AND THE TOOL

They are not going to be able to stay much longer. Old Yerok knows that the tribe is finished in this area. They will move on somewhere else, probably to a place owned by another tribe, and where the Tool is of no use at all.

He looks down at the clay model inside his shelter. It has taken him all his life to build, and now that life is almost finished it is becoming useless as well. The boxes, holes and chambers are an accurate reproduction of what has been found beneath the gravel and sand across the plain, but soon the whole thing – original and model – will be engulfed.

Every year the waters change. The rivers flow out of the ice wall and wash across the plain to the distant sea, splitting, crossing one another and rejoining, amongst the shifting pattern of gravel banks, sand bars and clay pans. They change their courses continually. This has always happened; the tribe is accustomed to it. Now, however, the ice wall is creeping out so far it is spreading over the plain itself.

Beneath the gravel, the sand and the clay, lies the Mystery. It was built by people a long time ago, and it was built to live in. Yerok can tell that by the pictures that he has found in it. Then it was destroyed by the sea, which he can tell by the layers of sand and mud that fill the rooms, chambers and passages, and the old seashells that cluster on the crumbling walls and the red powdery metalwork. Other people lived there afterwards, once the sea had retreated again, probably digging into it like his own tribe does. He can tell this by the skeletons piled in the mud layers above, that have to be shifted every time they dig downwards with the Tool.

The skeletons are of people, but of people quite unlike those of his tribe. His own people have longer arms and longer fingers and toes, as though they were designed to climb on things – rocks or even trees. Their teeth are bigger, as though they were meant to chew harder foods. Yerok feels a great sympathy with these old people, guessing that when he is dead, and that occurrence is not too far away, his skeleton will be found to be more like that of one of these ancient people than that of one of his own tribe.

He has known that for years, but of course nobody else noticed. He was born different, as if he were actually the son of a very distant ancestor, but one who had lain dormant, generation after generation, and only reappeared with Yerok's birth. His resulting greater intellect soon made him the leader of the tribe, and he led them into peaceful and plentiful times. It is his one great sadness that his children do not take after him: they are all the same long-armed, long-fingered, dull-witted, instinctively-acting creatures as their mothers.

He has always known there were riches to be found in the old dwelling places buried beneath the gravels of the plain. He built the Tool, and used it to dig into the sediments to find them. Now all the tribes within marching distance have drinking bowls, clothing and footwear, extracted from this plain by his tribe and traded for food.

Soon all that will be finished. The ice has been encroaching on the plain for as long as he can remember. In the gloom of his shelter he leans on his digging Tool and looks down at the meticulously-crafted clay model of the layout of the ancient dwellings – the model he uses to determine which part of the area the tribe should dig in next. Some of the places are gone already; those to one side have now been buried by the ice. The ice surge this coming winter will probably cover and obliterate the Mystery for ever.

Not only that, but the tribe is drifting apart. His two eldest sons, Hrut and Gultha, detest one another, both wanting to lead the tribe once he has gone. No amount of training will persuade them that it will be in the interest of all if they compromise. His death will be a sad blow for the tribe, and for all the other tribes in the area that benefited from the trading.

His death comes so suddenly that he has no time to recognize its approach. Hrut, silently behind him, brings down a rounded boulder from the gravel banks upon his head, and instantly obliterates the one force that has lifted

woodland-dweller
tundra-dweller
tropical tree-dweller
migrant
aquatics
memory people

Homo sapiens sapiens
Homo aquaticus
Homo caelestis
Homo sapiens machinadiumentum
Homo virgultis fabricatus
Homo glacis fabricatus
Homo silvis fabricatus
Homo campis fabricatus
Piscanthropus submarinus
Homo sapiens accessiomembrum
Homo mensproavodorum
Speluncanthropus
Moderator baiuli
Baiulus moderatorum
Homo dormitor
Homo vates
Alvearanthropus desertus
Homo nanus
Nananthropus parasitus
Penarius pinguis
Piscator longidigitus
Formifossor angustus
Acudens ferox
Harenanthropus longipis
Giganthropus arbrofagus
Arbranthropus lentus
Piscanthropus profundus

the tribe out of the surrounding savagery. The body that once held the last spark of civilization, a throwback to a sophistication that once was, falls limply into the clay plan of the ancient city, crushing the delicate walls and collapsing the whole intricate network.

With a cry of triumph Hrut grabs up the Tool. With this symbol he is now the master.

A shadow appears in the doorway of the shelter. It is Hrut's brother Gultha. Despite the slowness of his mind he sees instantly what has happened, and growls out a challenge. Hrut swings up the Tool in a wide arc, catching Gultha across the face and neck, and sending him staggering backwards to collapse bleeding on the gravel. He leaps out into the chill blue daylight and chops downwards with the Tool, until he is sure that Gultha is dead.

Then he stops to catch his breath. He is truly the leader now. He shakes the bloodstained trophy in the air in triumph – he has discovered the true function of the Tool.

5000 YEARS HENCE

TRANCER'S ESCAPE

He will be known as Trancer. He really has no name, since neither he nor his people have sophisticated speech, and so cannot think of themselves or of each other in terms of words. They have, however, a deep commitment and affection for members of their own group. Co-operation is necessary in the bleak mid-latitude tundra and coniferous forest where they live. To the north lie the snows and glaciers of the vast icecap; to the south, beyond the narrow belt of conifers, lies the vast sweep of cold steppe. There may be more habitable places beyond the chill grasslands, but they are too far away to contemplate.

The gnawing cold of winter is reaching downwards again, and the store of food that they have gathered this year is not very big. It will be difficult to feed all 20 of the group all through the winter, and impossible if they are raided by others.

Trancer is weary of fighting. Half of the food store in the shelter was gained by stealing from the other groups of the forest. This should not be. There should be plenty of food for everybody, and if there is not it should be shared equally. Certainly Trancer would be willing to share the mound of seed-cones that he is now carrying back to the shelter.

His weariness is temporarily overcome by a vague sense of achievement, as he is now carrying more cones than he has ever been able to before. He found the sloughed bark of a dying tree, and he kept piling cones onto it until it could hold no more. Then he carefully lifted it from the ground, and is now carrying the find, and the food, back to the shelter. If he had been using a thing like this all summer the whole group would have been able to gather much more food.

He breasts the edge of the narrow gully where the shelter is built, and begins to descend the slope carefully. Between the straight trunks of the trees the soil is loose – a yellow fragrant crust of decaying needle leaves and a rich black soil beneath. The shelter is a tightly-woven hood of sticks and branches, covered with a cosy layer of soil and needles. It is built half-way up the slope that faces the sun, so that it will be warmed by the earliest rays of next season, and yet is far enough above the floor of the gully to avoid the bitter frosts of the hollows. These hints for survival have been passed on by example from one generation to another.

Strong smells of crushed needles stop Trancer in his tracks. There is something wrong! He drops his load of cones and finds the shelter of an isolated bramble bush. Dimly, far along the slope, he can see a dozen figures heading silently towards the shelter. They are not his own people. It can only be a raiding party.

Trancer leaps from cover and runs down the cascading stream of soil and needles towards the shelter. He shouts to break up the raiders' stealth, causing the surprised faces of his group to appear at the entrance. Then, at the edge of his vision, he sees that the approaching party has abandoned its silence and burst out into a full-force attack.

The males of his group rush out of the entrance to defend the shelter, and Trancer turns to join them. Then he sees that the raiders are much more numerous than he had thought, and realizes that his little group is not going to stand much of a chance.

Wearily he steps back from the front line. He is not going to fight. He has had enough. He retreats into a corner of the shelter, closes his eyes tightly and curls himself up into a compact ball. With all his mind he wishes that this were over, that all the fighting were done and that the raiders had gone away. He wishes. He wishes!

He opens his eyes to a dark silence. Nothing is moving anywhere, and there is the unmistakable stench of death

A few of the swiftest woodland-dwellers have adapted to life in the tundra.

about him. His head aches, he is cold and, as he stretches from his cramped position, he finds himself to be unbelievably stiff. What has happened?

Slowly he crawls to the jagged shape of lightness that is the entrance to the shelter. Day is just dawning. He must have been asleep! In the midst of a battle! This could not have been an ordinary slumber.

As the sky lightens, he is able to take in what he sees about him. The raiders have left his tribe all dead; the bodies of his family are scattered limply around. They must have ignored him, thinking him dead as well. He does not look at the food store. He knows that it will have gone. He cannot possibly survive the winter now.

Then he looks closer at the bodies of his family. The spilled blood is dried to blackness, the faces are blue and sunken, the eyes have been taken by birds. These people have been dead for days!

He has been asleep for days! How can this be?

For the next few days and nights he can think of nothing else. His last recollection of the battle was of himself curling in a corner and wishing that it were all past. Now suddenly it *is* all past, as if he had wished himself into a temporary death to avoid danger.

If he can do that to avoid permanent death in a battle, could he also do it to avoid death by cold and starvation through a harsh winter?

It is worth a try. Best thing is to eat as much food as he can now – presumably his body will still need it while he is 'asleep', even though it will use it more slowly. Then he will have to find a comfortable sheltered place and wish that winter were all over.

He hopes that it will be as easy as that. It is his only chance of seeing the winter through until the warm growing times return.

SNATCH AND THE TUNDRA-DWELLER

This one will be referred to as Snatch. In shape, he is much like the generalized dim-witted temperate forest-dwellers generated in the laboratories of the now extinct genetic engineers 3000 years ago. He has the long body with the complex digestive system that allows him to eat almost anything, from leaves to grubs. His arms and fingers are long and nimble, but his legs are quite short – they were meant for pushing through thicket and undergrowth and for climbing the thick trunks of the deciduous trees, not for striding across the wobbly peat bogs and sharp grasses of the open tundra. Nevertheless the quickness of his actions has enabled him, and a few like him, to live on in his original area despite the fact that the landscape has changed from mixed woodland, through coniferous forest, to chill tundra bleakness in a few thousand years. Now an icecap sparkles on the northern horizon, where there was once the luxuriant green of forest in the time of his great-great-great grandfather. The standing waters of the peat bogs attract huge flocks of ducks and other birds for most of the year, and Snatch has become adept at catching these. By floating variously-shaped bits of wood on the surface of a pond he can entice the birds to land there. Then, when they are settled, he darts out of the concealing reed beds and grabs one before it can fly off.

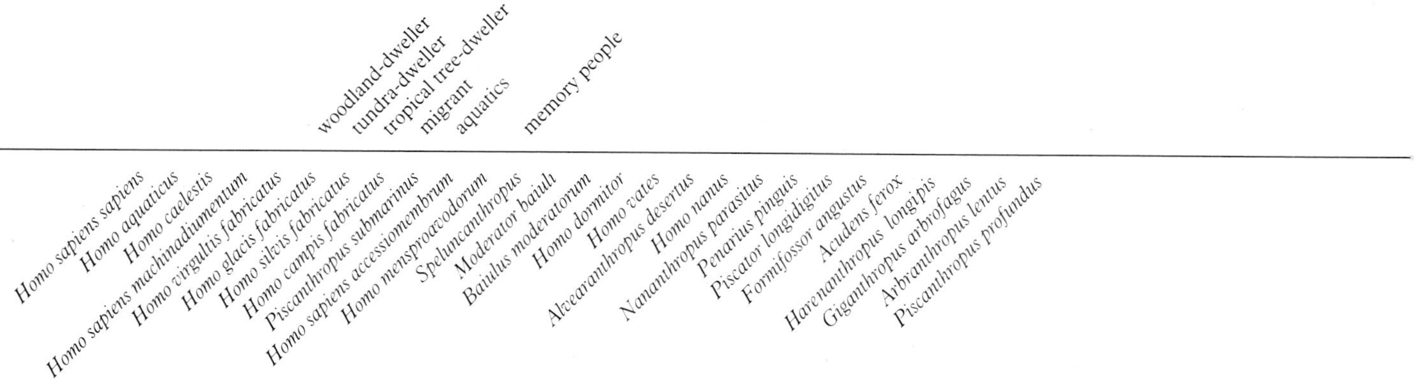

woodland-dweller
tundra-dweller
tropical tree-dweller
migrant
aquatics
memory people

Homo sapiens sapiens
Homo aquaticus
Homo caelestis
Homo sapiens machinadiumentum
Homo virgultis fabricatus
Homo glacis fabricatus
Homo silvis fabricatus
Homo campis fabricatus
Piscanthropus submarinus
Homo sapiens accessiomembrum
Homo mensproavodorum
Speluncanthropus
Moderator baiuli
Baiulus moderatorum
Homo dormitor
Homo vates
Alvearanthropus desertus
Homo nanus
Nananthropus parasitus
Penarius pinguis
Piscator longidigitus
Formifossor angustus
Acudens ferox
Harenanthropus longipis
Giganthropus arbrofagus
Arbranthropus lentus
Piscanthropus profundus

This time the weather has caught him out. The water of the lake is too cold for a long-term immersion, and the birds have not been coming. The sun is going down and the sky is about to turn to the misty purple he usually sees when he is almost back amongst his tribe; but this evening his tribe is a long, long way away.

Yet still he remains, reluctant to return empty-handed.

Over on the other side of the lake forages one of the tundra-dwellers, which also seems to be separated from its group. Its compact appearance, with its furry rolls of fat and its short arms and thick legs, makes it look as if it belongs in the landscape. It seems to be at home here, while Snatch, with his long limbs, does not. The two beings ignore one another. Their differing lifestyles do not put them in conflict, yet it seems to Snatch that the tundra-dweller should resent him, for being somewhere he does not belong; but he does not think about it too much. All he hopes for at the moment is that the other creature's movements do not interfere with his hunt.

Then, with a comical quacking noise, half-a-dozen birds settle on the still water, breaking up the reflection of the cold empty sky. Now Snatch squats into his hiding place amongst the fluffy heads of the grass, waiting for his chance.

It is a long time before any of the birds paddle close enough for an attack, but eventually they drift over towards his side of the lake. With a single dive, he throws himself out from the bank, his long arms and delicate fingers shooting out towards his prey. Startled ducks leap straight upwards from the water, flapping towards the sky and safety. One is too slow. The long fingers close around a webbed foot, and with a flurry of feathers it is dragged back as Snatch's body splashes downwards into the chill waters of the lake.

The numbing impact of the icy water cannot subdue Snatch's yell of triumph as he leaps out of the lake with his prize. Yet, before he has wrung the bird's neck, the chill has crept from his skin, through his flesh and to his bones. His newly-caught meat will be of no use to him if he freezes to death.

He rips the head off the bird, tears away the crop, and plunges his numbed fingers into the warmth of the carcass. It is not enough. He must find more body heat somewhere.

There is only one other big living thing nearby.

The tundra-dweller stands, still as a dead tree, watching all this with a dim curiosity. It shows no fear as Snatch approaches it carefully. Why should it? Tundra-dwellers have no natural enemies out here on the tundra, and no capacity for fear was ever designed into them by the genetic engineers all those millennia ago. For Snatch, there is a problem. How does he kill a big creature like this? His hands have only dealt with small mammals and birds up to now. The face, with the tiny eyes and broad nostrils, stares at him from within the frame of its voluminous neck ruff. There is no expression, and the creature does not flinch as Snatch drops his bird and throws himself at it, groping for a soft or vulnerable spot on its broad chest or its thick neck. Everywhere his fingers find tightly-matted hair and yielding blubber – nothing to hold or tear. Then, slowly, the great body leans over him and goes down onto its knees, pinning him to the springy vegetation. Snatch panics, and writhes and twists to pull himself out from under the mass of bouncing fat, but he is trapped. He can do nothing now but wait for the great creature to kill him.

After a while Snatch realizes that he is not dead. The tundra-dweller has not tried to kill him – it is just ignoring him. It went down onto the ground to reach Snatch's dropped bird, and is now eating it. Snatch was trapped by accident.

Night is falling, and it is warm in the folds of furry fat. As long as the tundra-dweller remains where it is, Snatch will survive; so he is quite happy to let it have his catch, in return for this life-saving imprisonment.

HRUSHA'S MEMORY

That way lies the end of the blizzard and the howling white blankness. Somewhere in that direction is the secluded dell of gentle green woodland, full of berries and nuts, with misty shafts of bright sunlight slanting through the leaves, bringing dappled patches of warmth, and the relaxed noises of chattering, twittering birds heard over the gurgling of a little brook as it splashes over the moss-covered rocks.

How does Hrusha know that? She has never been here before. She has never even seen a gentle green woodland, would not recognize berries and nuts for what they were, and would be alarmed at the strange noises of twittering birds. Yet somehow she knows that these things are to be found in the direction in which she is walking.

Her colony by the seashore is starving. The colder weather this year has meant that fewer fish have come to the beaches, and fewer herbs are growing along the spume-blown shingle that separates the grey ocean from the white

5000 YEARS HENCE

MEMORY PEOPLE

Homo mensproavodorum

As the genetic engineers have long gone, there can be no further artificial changes. When climates and conditions shift, altering habitats, the inhabitants must normally adapt or evolve to survive. But the woodland-dwellers have a different option.

A genetically-manipulated but latent ability to recall the long-term past is forced to the surface by climatic extremes. A group of Homo virgultis fabricatus *become the memory people.*

of the icecap. Others have travelled out from the colony both ways along the coast, to try to find new sources of food; but few have returned, and those who did come back reported no success.

Now Hrusha and her mate Vass have tried going inland instead: a bold and dangerous choice, and one that Vass is constantly regretting. Inland is nothing but snow and ice.

As they trudge onwards the blizzard develops, intensifies and turns everything to a featureless whiteness. Their vision is blocked by the relentless glare, their hearing muffled by the unchanging howl of the wind, and their sense of touch numbed by the cold.

Suddenly, with her normal senses dulled by the disorientating surge of the blizzard, Hrusha remembers something that she could not possibly have experienced, and with excited gestures urges Vass to follow her. This is too much for her mate, who turns and tries to find their tracks, hoping to follow them and make his own way back to the coast.

Acting on the hunch that is stronger than her mating bond, she trudges in the direction her senses dictate, deeper and deeper into the blasting, blinding blizzard, and suddenly the snow gives way beneath her. She falls, tumbling with the snowy lumps, and ends up face down in a shallow drift. As she struggles free she finds that the wind has dropped, and she is lying in a sheltered ice-free valley. Dark rocks jut from black frozen soil, and an ice-bound stream winds along the valley floor. The most remarkable features of the landscape, though, are the hulks of dead trees, standing black and branchless, frozen and upright, where they died of cold an unimaginable time ago.

This is the green and leafy dell that she remembers, but changed by time and creeping coldness. How can she remember this, when the trees she sees around her have obviously been dead since the time of her father's father's father's father? Could that be it? Could the landscape have been seen by one of her fathers? Could the memory have been passed on to her, like her distinctive hair and eyes? As far as she knows, none of the others of the colony have had that experience before. Certainly her mate Vass has not.

She settles by the frozen stream, smashes the thin covering of ice, and drinks from the cold water beneath. Surely this experience could be useful. Surely she must be able to remember other things that her ancestors saw and knew – things that would help the colony in its time of trouble. She must think.

Where is there food?

Where the stream comes out, comes the answer, in a lake full of fish, a lake that never freezes over even in the harshest of winters. She remembers that now.

Weary from her journey, but now filled with hope, Hrusha rises and walks heavily down the frozen soil of the valley following the winding stream between the dark rocky banks. Eventually the valley gives out and a plain stretches out before her. The blizzard has abated and she can now see for some distance. In the middle of the plain is a white expanse of perfectly flat snow that can only be the lake. It is frozen now, but the ice is quite thin, and it seems very likely that fish still live in there.

That is what the colony needs to know. She turns to retrace her journey to the coast, and there in the distance she sees a figure coming towards her, a figure she seems to recognize. It is not Vass, is it? No. Vass does not have the knowledge that brought her here. It must be someone else who can remember this place from long before they were born. Someone else who has the ability – an ability forced to the surface by the jeopardy of the colony. The figure is closer now, and she sees that it is Kroff, the son of her cousin, a person she has always ignored since the two of them have never had anything in common.

That must change now. If Kroff has the knowledge, then he is a far more suitable mate for her than Vass ever was. This needs to be seriously considered.

TROPICAL TREE-DWELLERS

Plenty of fruit is available in the tropical treetops, so there is nothing to worry about here. Like the extinct monkeys and apes, the tree-dweller (he has not the wit to consider himself as an individual let alone as a being with a name) climbs the vertical trunk through the luminous green of the leafy canopy, and scampers four-footed along a broad bough, forking on to a thinner branch and finally along slender waving twigs to reach the point where the bunches of fruit dangle invitingly. Hanging upside down now, he reaches outwards with his narrow prehensile fingers and delicately prises the bunch free from its stalk. Some fruits drop off, falling with a fading 'plop, plop' through the layers of leaves and twigs below, away to the forest depths. These are immediately forgotten, as he has secured enough for his needs.

This is his whole life. It is of no relevance to him that the equatorial tropical forest belt of the Earth is narrower now than it has been at any time within the last million years,

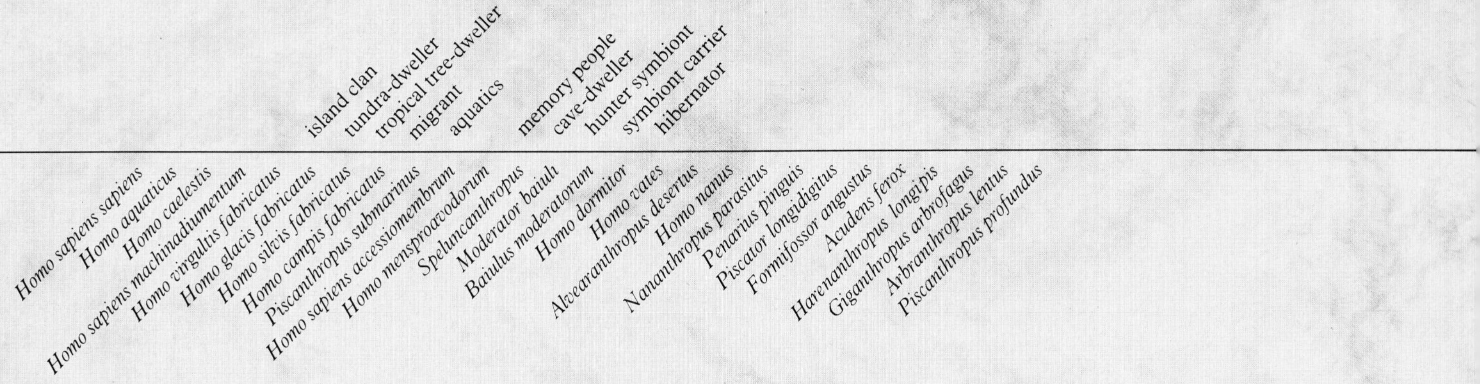

island clan
tundra-dweller
tropical tree-dweller
migrant
aquatics
memory people
cave-dweller
hunter symbiont
symbiont carrier
hibernator

Homo sapiens sapiens
Homo aquaticus
Homo caelestis
Homo sapiens machinadiumentum
Homo virgultis fabricatus
Homo glacis fabricatus
Homo silvis fabricatus
Homo campis fabricatus
Piscanthropus submarinus
Homo sapiens accessiomembrum
Homo mensproavodorum
Speluncanthropus
Moderator baiuli
Baiulus moderatorum
Homo dormitor
Homo vates
Alvearanthropus desertus
Homo nanus
Nananthropus parasitus
Penarius pinguis
Piscator longidigitus
Formifossor angustus
Acudens ferox
Harenanthropus longipis
Gigantanthropus arbrofagus
Arbranthropus lentus
Piscanthropus profundus

that the cooler climates have been encroaching from the north and the south, bringing their windy grasslands and barren deserts with them. The only significance to him is the fact that when he is in the gloom of the lower branches he often sees, on the forest floor, bands of strange creatures moving purposefully in a particular direction. Since he rarely ventures down onto the floor anyway, he just ignores them.

The lost fruits, dented and bruised by their fall through the branches, at last thump softly down into the decaying plant matter of the forest soil. A group of gaunt long-legged plains-dwellers, uneasy and out of place in this strange environment, but driven from their grasslands by increasing cold and ravening packs of wild creatures, starts at the sudden noise. Then, when they see the fruit that has fallen, all four of them pounce upon it, scratching and tearing at one another in their attempts to reach it first.

This drama is completely irrelevant to the tree-dweller. There is always plenty to eat up in the sunny heights and he can leave the lower shades to those strange beings.

It is in the far north and the far south that the ice age is causing its havoc. Fluctuating icesheets and glaciers, together with unstable weather patterns, are forcing high- and middle-latitude inhabitants to resort to drastic measures and changes in lifestyle just in order to continue living, and encouraging genetic changes in body and mind that could not have endured if the environment had remained constant and unchanging. Here, in the tropical forest, however, things have not altered for thousands of years. The tree-dwellers have a constant supply of fruit and insects in their leafy canopies, so there is no need for them to move to new areas or to change in any way.

10,000 YEARS HENCE

SYMBIONTS

The symbionts are marching.

A temporary and small retreat of the northern icecap has created vast new tundra areas over the northern continents. For the first time in 5000 years the rate of melting of the edge of the glaciers is exceeding their rate of southward movement. In effect, the edge of the icecap is melting back. Rocky debris, broken up by the weight of the ice and shoved along the ground by the southward movement, now lies in hummocks and thick beds of mixed clay and boulders. Here and there a long winding esker (a steep ridge of rubble marking the old course of a subglacial river) snakes across the plain. Huge lumps of abandoned ice embedded in the clay melt slowly, gradually becoming lakes.

Yet below the ice-free surface the soil is still permanently frozen. Little grows here, except for the hardy grasses and reeds along the sides of the lakes, and the mosses, lichens and heathers that form tussocks over the rocky soil. Away to the south lie the great forests, which are already spreading northwards into this newly-exposed land, with their outposts of stunted willows, birches and rowans, backed by the dark palisades of spruce and pine. It will be a short-lived advance if the ice moves south once more.

It is the domain of the symbionts. From a distance as they trek across the plain they look like the tundra-dwellers of old, but they seem to be bigger and rather top-heavy. A closer look shows them each to have two heads – a large one surrounded by the woolly ruff of blubber, with small eyes and large nostrils, and beneath the chin a smaller head with big ears and active, darting eyes. The herd consists of about 30 individuals, adult and juvenile. They follow the biggest, whose lower head seems to be looking around all the time for the best way to travel.

It stops, staring away into the distance. A dark flock of birds circles in the far sky, something that should be investigated. The leader's arm shoots out in that direction – an amazingly slender arm for a creature of such a size – and it turns towards the distant flock. The rest of the herd turn as well, each one also shooting out an arm.

After a while they come to the site of the disturbance. Most of the birds are hawks, and every now and again one swoops to the ground and carries away something small and furry. There are small foxes there as well, but these turn and scatter as the symbionts approach. The cause of the activity is now visible before them. A mass of small rodents – lemmings – is on the move. Every now and again, in times of relative plenty, they breed prodigiously, until there are so many that the food in their area runs out. Then they move *en masse* to find new foraging regions. The symbionts have just come upon one such migration, a moving furry layer that stretches in a straight line along the ground towards a possible distant food source.

If the movement of the rodents is remarkable, what then happens to the symbionts is even more strange: about half of the individuals fall apart, literally. Each one resolves

10,000 YEARS HENCE

SYMBIONT CARRIER

Baiulus moderatorum

Two species form a single unit of value to both – symbiosis. The woodland-dwellers have skills that their carriers lack. The hunting ability of the swift forest-dweller provides enough food both for itself and its slow-moving carrier. The tundra-dweller, in turn, provides both with general movement and protection against the cold.

Lacking thick fur and insulating layers of fat, Moderator baiuli *can only hunt in short bursts before needing to return to the body heat of its carrier. Communication is by touch.*

itself into two separate creatures. The huge hairy arms of the tundra-dweller that were clutched across its chest open up like doors, and to the ground drops a spindly figure – the owner of the second head and the pointing arm. The slimly-built creatures are running as soon as they hit the ground, and ten of them plunge into the moving mass of lemmings, snatching and killing as they go. The remaining symbionts, mostly females and young, stand watching, shouting encouragement in words and noises that only members of their own group can understand. The tundra-dweller shapes vacated by the hunters stand immobile and silent.

After a while the hunters gather up the rodents that they have killed and bring them back to the group. They are handed around to the figures clutched to the bosoms of the tundra-dwellers. Then each hunter returns to its own tundra-dweller/carrier and, with a touch and a word, it is gathered up into the great arms. For a while they eat. Each lemming is partly eaten by the being at the symbiont's chest, but then the greater part of it is handed up to the great mouth in the head above. The tundra-dweller part of the creature receives the food passively, and eats it all.

This strange state of affairs began thousands of years ago. When the hunters (the humans that had been engineered to live in the temperate forests) spread out to hunt on the tundra, with the coming of the current ice age they had to adopt all kinds of strategies to keep themselves warm and to survive. Some found that they could live close to the dull tundra-dwellers and share their body heat. The tundra-dwellers did not mind this, if the hunters shared their food with them. So the symbiotic situation gradually developed, until now the hunters could not travel by themselves in the tundra and these particular tundra-dwellers would not be able to survive on their own.

Once the food is eaten, the group sets off again. The smaller hunters can talk to one another, using a simple language, but each communicates with its tundra-dweller/carrier by nudges and gestures – a pointing of the arm is enough to tell the carrier to go, and which way. They follow the lemming march, as there will be good eating here for a day or two.

HIBERNATORS

The trance-like state is not now as deep as it as a few days ago. The warmth of spring is filtering through the cocoon's insulating layers of fibre and wood, registering slowly on the dulled nerves and sensory system of the sleeper, triggering a slow increase in his metabolism and bringing consciousness nearer. His mind emerges from total blankness into a dream state, in which he relives and consolidates the hunting and gathering techniques that he learned last season. In his dream he sees the forest of his home, firstly as it was when he was a child, then as it was more recently. The most recent memory-dream is of his mate of last season, and hopefully of this season also, and the thought of her excites him so much that the final barrier of conciousness is broken, and he is awake.

With a groan of momentary disappointment that the last vision was only a dream, he stretches himself, pulling open his eyelids against the mucus that gums them together, and unfolding his limbs which are so stiff that they almost creak. With a struggle he breaks through the covering of vegetable matter and into the spring smells of the coniferous forest.

The spring flowers – the gentians, orchids and saxifrages – are out and blooming, as they are whenever he awakes from his winter sleep, but the sun is low in the sky. Spring is early this year, therefore the climate must be becoming warmer.

Then the pain of hunger strikes him and he digs at the remains of his food store. Several times throughout the dark winter he broke his trance to feed and now there is little left. Most of the tubers have rotted and the seeds germinated, but there are still a number of items that are edible. These he devours with no hesitation, to give him the strength to look for more.

There is plenty of food about, since it is the beginning of the time of swarming insects and the damp soil and decaying needles underfoot house a vast array of luscious wriggling things. Beneath the bark of the trees, too, grubs and beetles burrow and tunnel in their millions, and birds are here as well, having travelled up from the south, as his mate will hopefully do, to feed on the insects. When his stiffness has worn off, and he has built up his strength again he will also be able to catch the birds and the little rodents that have come out to feed on the tender shoots and saplings.

Looking for food, he rips the bark from a fallen tree, one which must have died during the winter. He remembers when it was merely a sapling – over 60 years ago, but numbers mean nothing to him. He only remembers.

After building up his strength for a few days he sets about the task of building his fortress. It will be made of wood, comfortable and soft inside, but harsh, jagged and

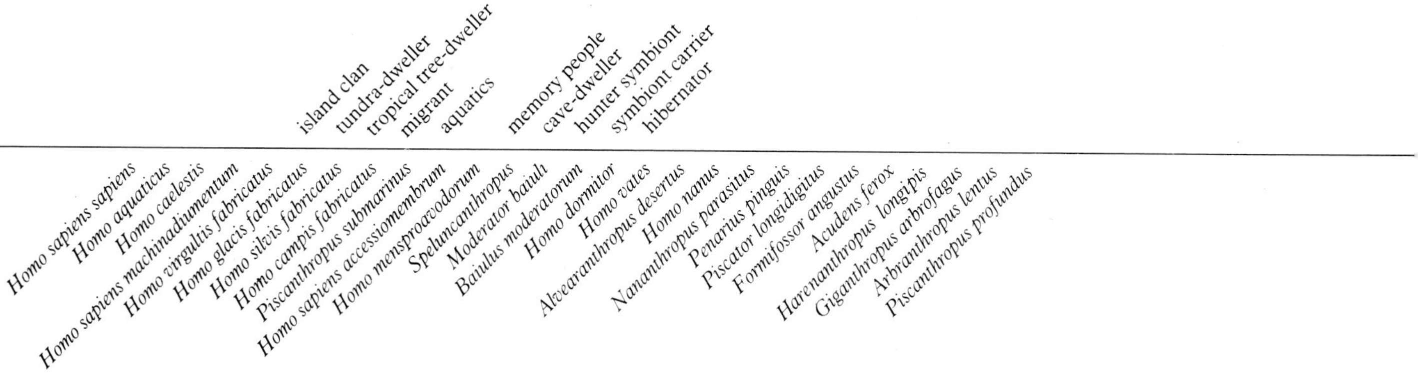

island clan
tundra-dweller
tropical tree-dweller
migrant
aquatics
memory people
cave-dweller
hunter symbiont
symbiont carrier
hibernator

Homo sapiens sapiens
Homo aquaticus
Homo caelestis
Homo sapiens machinadiumentum
Homo virgultis fabricatus
Homo glacis fabricatus
Homo silvis fabricatus
Homo campis fabricatus
Piscanthropus submarinus
Homo sapiens accessiomembrum
Homo mensproavodorum
Speluncanthropus
Moderator baiuli
Baiulus moderatorum
Homo dormitor
Homo vates
Alvearanthropus desertus
Homo nanus
Nananthropus parasitus
Penarius pinguis
Piscator longidigitus
Formifossor angustus
Acudens ferox
Harenanthropus longipis
Giganthropus arbrofagus
Arbranthropus lentus
Piscanthropus profundus

defensive outside. It needs to be, since there are many marauding males about that would fight him to death for a fertile female like his. He builds his fortress on what remains of last season's, and that is quite a lot. As the years go by his building techniques improve and his structures become more durable.

Little remains, however, of the guide walls, and these have to be rebuilt every spring. Reaching out in two directions in a huge V-shape, open end to the south and with the fortress at the apex, the structure stretches for over 2000 paces in each direction. It is made of sticks pushed into the ground and thinner sticks woven in between. It is not meant to be a barrier, but more of a marker across the landscape. His mate has wintered in the milder climates away to the south, and will be travelling northwards very soon. It is essential that she does not miss the fortress and go blundering on northwards, or end up in some other male's domain.

With construction completed, he starts to build up the food supplies in the fortress itself. After a few days he hears an excited chatter, and he looks expectantly from the mouth of the now comfortable fortress. She is there, walking confidently up the side of the barrier.

Yes, she carries the winter's baby with her.

With joy, one of the few emotions he can feel, he rushes to meet them, and to fondle her and stroke the child he sees for the first time. A female. That is good: there are enough males around. This is the first child that he has had by this female, although he has had many others by other mates.

Females are much shorter-lived than males. They cannot sleep the cold times away, as they have to travel south to give birth in the winter. Many of his females aged and died during his life, while many others became lost in the migration, dying on the trek or ending up in other fortresses.

Each creature has its allotted life span. Barring accident or disease it survives for about 2000 million heartbeats. For the migrating females these heartbeats average about 70 per minute. For the hibernating male this average is kept up during waking times, but during the late autumn, winter and early spring it drops to about 20 per minute. The remainder of his bodily functions slow down accordingly. As a result the male's lifespan is between four and five times the length of the female's.

In the dimness of his weak imagination he sometimes thinks that it would be better if babies were born during the summer so that they could all hibernate together; but this would not be possible unless the growth of a baby inside the female could be speeded up or slowed down, so that the offspring of the spring mating emerged at a more convenient time.

That cannot be... yet.

LEADER OF THE CLAN

There is no way across the water any more. In times past, low water exposed broad expanses of brown rippled mud, with winding glistening creeks, joining the flat marshes of the clan's domain and the infinite woodlands of the country beyond. At these times the clan could squelch across the mud, churning up the black stinking subsurface, and go on short forays amongst the trees and forests of the mainland.

That is no longer possible for the mud flats are now permanently submerged. The clan can know nothing of the reason, the shrinking icecap thousands of kilometres to the north. They would not understand that the melting ice is pouring water back into the oceans, and that the sea levels are rising worldwide. They only realize that the island on which they now live is more isolated than it has ever been at any time they can remember.

It does not disturb them. The woods and marshes of the island supply plenty of food for the small numbers that live there, and the rainy climate provides enough drinking water. It has been only occasionally, in times of severe hardship, that any of them have actually crossed the mud to the mainland to forage. Mathematics and measurement do not come into their lifestyle at all, but if they did they would realize that the 200 square kilometres of the island are just able to support the 20 individuals of the clan.

The leader thinks of it in another way. He can walk right across the island in half a day. This walk takes him through bushes with fruit and undergrowth with tubers, and between the trunks of trees with nuts. Everywhere there are birds and small animals that can be caught. Walking around the coast takes three days of daylight, and takes him across beaches with burrowing creatures, over rocks with attached shellfish, and through saltmarshes full of birds. A clan the size of his is well supplied, for the moment.

There were times when food was short, and they all went onto the mainland; but that has always been dangerous. Other clans live there, and they do not take to strangers. Now they will have to deal with any shortages in some other way.

Certainly it will be best if the clan does not grow any

more. More mouths to feed will be a disaster. If they can all eat less as well, it will help. The leader cannot anticipate any of this. His whole attention is taken up in ensuring that all his people have enough food. He has noticed, however, that one of his daughters, a very big-boned and heavy female, becomes hungry and ill more often these days. At the same time another of his daughters, this one very much smaller and more lightly-built than her sisters and brothers, has a small appetite and is the healthiest of the whole clan. She will certainly reach maturity and breed.

DISAPPEARANCE OF THE PLAINS

Rain falls. It now falls for long periods and the grasslands are losing their character. Instead of one short rainy season in the year followed by long periods of dryness, there is now more rain all year round.

The grasses thrived under the old conditions. Their tops were shrivelled off by the sun, grazed away by animals and burned by periodic bush fires, but they survived because of their protected underground stems, and grew again from ground level. Few trees or bushes flourished under these conditions, but the plains dwellers also did well here. Their exclusive diet of grass meant that they could live here where no other large creature lived. They could spend the dry seasons in the thorn thickets that bordered the grasslands and separated them from the humid tropical forests of the equator, and they migrated out over the grasslands proper during the wet season, feeding as they went. Other large creatures could not cope with this existence.

Now, with the more frequent rains, the thorn forest is spreading over the plains, and trees are growing where once there was only grass. With the new conditions different creatures, ones that hunt meat for food, are creeping out of the tropical forests. More and more often the plains dwellers have to take themselves off out of danger. With their immensely long legs they can quickly outpace any enemy, but this is becoming more and more frequent. It is wasting a great deal of energy and eroding valuable eating time.

Over the past few thousand years the plains-dwellers have faced problems like this, many times. Sometimes, when it seemed as if the grasslands were going to disappear, herds of them went through the thorn thickets and into the depths of the great rainforests, in the hope of finding new pasture. None ever returned. Few went the other way, where the grasses became shorter and sparser, where food became harder and harder to find, and where even small creatures became rarer and more difficult to see. The grasses in this direction eventually gave way to rocky and sandy wastes, where the rainy season was even shorter and less reliable than it was on the plains. In these previous times of crisis, however, the problem was never long-lived: the grasslands established themselves once more.

Now, with the increased rainfall, the grasses as the plains-dwellers knew them are becoming obliterated by thorn forest. The only reliable expanses of grass seem now to be found in the once-desert areas, and even these unbeckoning wastes are changing because of the increased moisture. Grasses and low plants are finding purchase in the harsh rocky soil that once they found uninhabitable. Perhaps in this direction lies the future home of the lanky plains-dwellers.

CAVE-DWELLERS

It has been, after all, just another temporary respite. The cold weather returns. Winter becomes long and bitter, while summer dwindles into the briefest of seasons, unable to melt the snows deposited the winter before. The southward movement of glaciers is again faster than the northward melting of their snouts, and the icesheets spread into the plains and lowlands of much of the northern hemisphere.

He seems to retire into his hibernation earlier and earlier each year, and his sleep lasts longer and longer. At least the fish have still been coming to the stream outside his sleeping cave. There was always food available for him in the narrow valley.

This year, however, it is different. After he awakes, he can hardly approach the entrance of the cave, so bright is the glare of the snow outside. He waits for night to fall, so that the outside light will not hurt his eyes after his long slumber. In his hunger he chews the moss from the cave walls and the fungus from the floor. After a while the light fades, and he prepares to face the outer cold. Suddenly, there is a strange screeching noise from deep within the cave behind him. It becomes louder and, with a flurry of wings, a black flock of bats hurtles upwards from the depths and out through the cave mouth. In a reflex, stiff from long hibernation but still good enough for the purpose, he shoots out his arm and grabs one of the furry creatures from the air. It squeals once as it dies, and he eats it whole, chopping up the body with his sharp front teeth

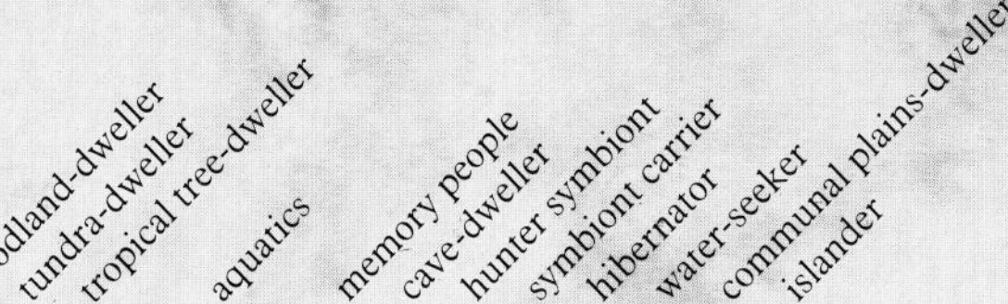

Homo sapiens sapiens
Homo aquaticus
Homo caelestis
Homo sapiens machinadiumentum
Homo virgultis fabricatus
Homo glacis fabricatus
Homo silvis fabricatus
Homo campis fabricatus
Piscanthropus submarinus
Homo sapiens accessiomembrum
Homo mensproavodorum
Speluncanthropus
Moderator baiuli
Baiulus moderatorum
Homo dormitor
Homo vates
Alvearanthropus desertus
Homo nanus
Nananthropus parasitus
Penarius pinguis
Piscator longidigitus
Formifossor angustus
Acudens ferox
Harenanthropus longipis
Giganthropus arbrofagus
Arbranthropus lentus
Piscanthropus profundus

and grinding up the little bones with the massive molars at the back. The warm blood and juices warm his inside, and presently he begins to feel fully awake. The torrent of bats is still blasting out of the cave mouth, and he grabs another to eat.

There is now no need to go outside. The humble plants at the cave entrance and the unending supply of bats could keep him alive here for ever. Then he remembers that there are birds that nest here, too, high up in the cracks and gullies of the cavern walls, and small shrimps and insects in the running waters deep down. These will be good for eating as well. He does not need to go outside in the cold, at least not tonight. He turns his back on the grey entrance and begins to grope his way back down the tunnels into the comfortable depths.

Dimly he wonders if any others of his type realize how much food there is to be had down here. Sometime he will go out into the chill and find some of them and bring them down.

Another time. Tonight he has food to find.

50,000 YEARS HENCE

FAMILIES OF PLAINS-DWELLERS

The harsh hot wind hums over the wispy grass and red-hard soil of the semi-desert, drying out the skin of any creature exposed to it. Climates are changing again and the whole of the world is feeling the effect. Here, the grassland that had once been desert is turning back to desert again. After 40,000 years in which the climate has been relatively settled, in which seasonal rains have been enough to sustain sufficient vegetation for the herds of plain-dwellers, the food chain is becoming unstable once more.

Over the years the plains-dwellers have changed. They still subsist largely on the tough grasses, but now they have begun to vary their diet and lifestyle in a number of ways. They have given up their wandering life and now stay in the few places where they know there is water. Their broad hands, with the blade-like callouses along the edge, have proved to be ideal for digging in the ground, something that was first discovered when they had to dig for water in the cracked and sun-baked hollows that in the rainy season form the muddy water-holes. Soon it was realized that food, as well as water, exists below the surface. Now they dig frequently for the moist tubers and underground stems that keep many of the desert plants alive during the dryness. Occasionally they will also chew and swallow a large insect, or a burrowing mammal or lizard, but these are invariably thrown up and spat out in disgust. The plant-eating digestive system with its bacterial vats is far removed from the omnivorous stomach and intestines of the plains-dweller's far ancestors.

Turning his back on the scorching wind, the plains-dweller heads back towards the oasis with his load of tubers. Intermeshing his long fingers makes a kind of basket of his hands, and this can hold a large quantity of food. Now he must guard them against any enemies, for there are other groups of plains-dwellers around, and they would stop at nothing to get at somebody else's food store. It is not just the dry wind and harsh sunlight that are enemies to the plains-dwellers – they must fear members of their own kind. Not their own family, however; the tough conditions ensure that every family is tightly bound and co-operative.

There is trouble back at the oasis – he can feel it as soon as he crosses the rocky ridge and descends into the hollow. Home is there, as safe and impregnable as usual, its red baked-clay walls rising like cliffs, with its entrances guarded by his young brothers and sisters; but there is an air of strife about the place. It does not feel as secure and home-like as it normally does. He passes through the entrance with no trouble – his brothers and sisters recognize him instantly, and he enters the shaded courtyard within. Over by the well there seems to be some kind of dispute. He ignores it for the time being. His first duty is to store the food that he has brought, and he does this in one of the cool storage cells dug into the hard clay soil. Then he emerges and goes to the well to see what is happening.

It is the usual trouble. One of the younger females, his older sister, has been caught mating. Their mother is understandably enraged, as she is the one who gives birth in this family. The turn will come for the other females when she has become barren, or is dead and gone, but that will not be for a long time yet. Meanwhile the sons and daughters must concentrate on what they have to do to keep the family alive, and not waste their precious time in irrelevant mating. Conditions are too harsh for this. Everyone must do his or her duty, continually, if the family is to survive. There can only be one female giving birth at one time, and she must have the whole-hearted support of everyone. Otherwise the birth-rate will run away, bringing

Without some water, no species can survive. The descendant of woodland-dwellers, Homo vates *has retreated north before the advancing desert. Now he can retreat no more. He must find water or die.*

the family number beyond the present viable level of 20, and the family will collapse through lack of resources.

His sister seems abashed. She knows what she has done. It appears that when she was confronted with her crime she turned on their mother and attacked her, evidently in some kind of half-hearted bid to oust her from her breeding position; but the mother is not yet old enough or frail enough for anything like that. Now his sister, bleeding from cuts to the face and shoulders inflicted by their mother's hand-blades, scuttles through the crowd to the entrance of Home. She will never be welcome here again. Already her brothers and sisters are picking up stones to see her on her way. They will be sad to lose her. He duties as a wet-nurse will be missed, but not for long since some of the younger sisters are almost old enough. It is better, on the whole, for the family to lose an unreliable member.

Outside the entrance she stops and looks back. The first stone is cast, and misses. The second hits, but she does not go. Outside she will die, unless the older brother who mated with her comes out to join her. Then they may go far away and possibly found another family, if any of the other families will let them.

The brother is not coming. He has realized his error and will stay, doing his duty to his mother. The sister eventually realizes this and, still bleeding from cuts and bruised by the stones, walks off into the barrenness to die.

The family will survive.

THE ADVANCING DESERT

He was no plains-dweller adapted to the searing heat of the desert. He was in no way prepared for the dryness that had killed his tribe and was now killing him. His dark skin was protecting him from the worst of the sun's blast, but without water he was going to perish.

They could not move northwards any more, his tribe and he, despite the fact that the arid lands were moving northwards year by year. They had tried to stay ahead of it, keeping abreast of the zone where there were still enough trees to supply their fruit and seeds, and still enough small animals for their protein; but now the people of the lush north barred their path. They were not moving away from their homelands just because the people from the marginal lands needed to survive. After a particularly fierce battle the southerners had to retreat and find their own way of life in the desert.

It has not worked. They are all dead but one, and he has not much longer to go.

The sun in his eyes dazzles him, the singing of the sands dulls his hearing, the dust in his nostrils clogs his sense of smell and taste. He is wandering lost and without the help of any of his senses. Hallucinations about his tribe force themselves upon him – waking nightmares that chide him for surviving while the rest perished. No matter, he is about to join them.

Then comes the other hallucination; the one about the water. Over there, about 500 paces away, if he has the strength, and just below the soil surface beneath the rocky ledge of a gully, lies enough water to save him. It is only a dream and not worth any attention.

Yet it is not like a dream, but more of a conviction that has been put into mind pictures. Over there lies enough water to save his life. He does not imagine it, he *knows* it.

He finds the strength to pull himself in that direction, slowly, on hands and knees against the abrasive sand and rock, until eventually he sees ahead of him the rocky outcrop and gully of his hallucination. With a final burst of energy he pulls himself into the hollow, and begins to dig the loose soil. After a while the fine powdery sand becomes coarser, cooler and more cohesive. It is coming out as

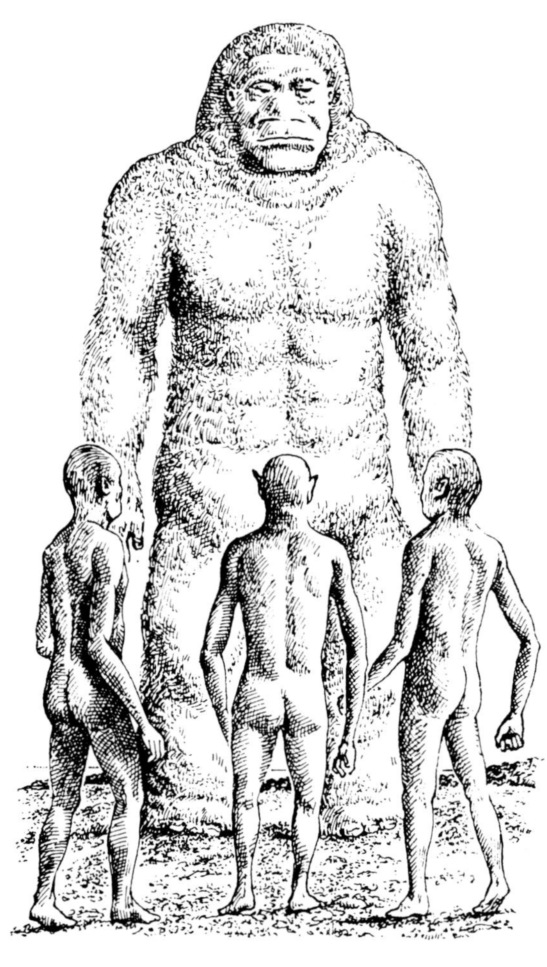

Isolated from mainland evolution, island-dwellers have developed a high-protein diet and reduced in size. Now, as a new species Homo nanus, *the islanders return to the mainland, where the tundra-dwellers have adapted as a leaf-eating forest people.*

lumps, stuck together by moisture.

He crams a handful into his mouth and sucks the water from it. Then he digs further and finds the sand becoming wetter and wetter.

After a long time he is finally refreshed. He must now look for food, which is another difficulty; but there will be plants and small burrowing animals around. Somehow he has solved the main problem of living in the desert.

He can see water.

ISLANDERS

The icecaps and glaciers are in full retreat now, melting away to the poles and withdrawing up the mountains. The climate is becoming warmer, changing the conditions not only in the arid tropics but over every climatic and vegetation zone of every continent. The retreat of the ice changes not only the climate but the geography as well. Meltwater, gushing from the rounded ice-tunnels and widening crevasses, floods into intertwined rivers that wash across the gravel plains and empty into the ocean, causing sea levels to rise over the whole world. In some places, however, once the unimaginable weight of ice is removed, the land surface rebounds like a slow spring, lifting it above its former level, and causing the sea level to fall back. Then there is the volcanic activity, mostly at the edges of continents and in strings of islands arcing across the oceans, producing new lands and destroying others.

All in all, it is a time of appearing and disappearing islands, of continents joined by land bridges which then submerge, and of lowlands engulfed by the seas and shallow seas that become plains bounded by the banked shingle and sand of former beaches.

The islanders have always found it easy to move from island to island, floating upon the trunks of trees wrenched from their forest stands, or on rafts built from the stems of smaller trees lashed together by vines and creepers. They have used vessels like this to support them while they dived for fish in the straits of the archipelagos. Now, however, this activity is dangerous. The changing weather patterns are producing unfamiliar winds and frequent storms, and changing the sea currents between the islands. More than one raft of island voyagers has disappeared in recent memory.

One has found itself on the beach of the mainland – a region the existence of which was only guessed at by the island people. After the rigours of the accidental voyage the new country may be either an unending source of plenty to the small hungry group of five islanders or deceptively barren. The islanders' original digestive systems allowed them to eat almost anything, but millennia of island-dwelling on crags and slopes that supported few nutritious plants have changed all that. Now they can only subsist on the high-protein diet that they gained from birds and their eggs, and the fish and shellfish of the sea. No birds seem to nest on accessible crags here, and the shingle beach gives little purchase for shellfish.

There may also be enemies. Some huge figures are moving about down the beach. In build, they are somewhat like the islanders, but they are more than twice their size, and very slow-moving. There are about ten of them.

The islanders do not know these creatures for descendants of the tundra-dwellers. The tundra is dwindling away now, but for many thousands of years groups of its inhabitants have been spreading southwards, changing their diet and adapting their lifestyle as they went, through the coniferous forests and into the zone of deciduous woodlands. Because they have been forced to change all the time they have a better chance of survival than the groups of their relatives who remain static on the tundra. Now they are massive leaf-eating forest-dwellers – dim of wit but quite adaptable to changing conditions. However, they do retain the thick deposits of fat that are now superfluous to their purposes, and indeed could be disadvantageous to them in the hot times that may come. Nor do the islanders realize

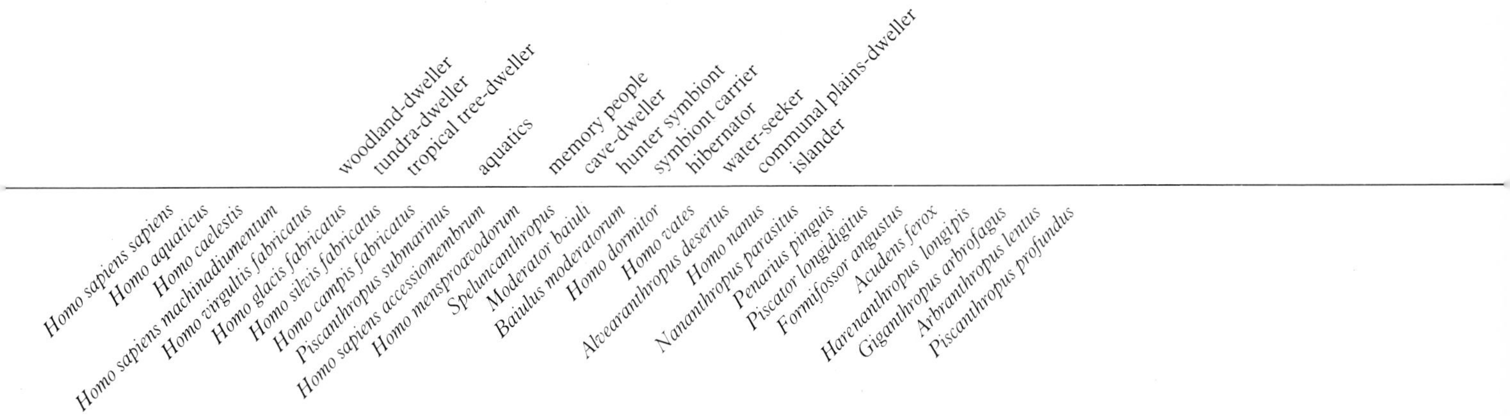

woodland-dweller
tundra-dweller
tropical tree-dweller
aquatics
memory people
cave-dweller
hunter symbiont
symbiont carrier
hibernator
water-seeker
communal plains-dweller
islander

Homo sapiens sapiens
Homo aquaticus
Homo caelestis
Homo sapiens machinadiumentum
Homo virgultis fabricatus
Homo glacis fabricatus
Homo silvis fabricatus
Homo campis fabricatus
Piscanthropus submarinus
Homo sapiens accessiomembrum
Homo mensproaevodorum
Speluncanthropus
Moderator bajuli
Bajulus moderatorum
Homo dormitor
Homo vates
Alveearanthropus desertus
Homo nanus
Nananthropus parasitus
Penarius pinguis
Piscator longidigitus
Formifossor angustus
Acudens ferox
Harenanthropus longipis
Giganthropus arbofagus
Arbranthropus lentus
Piscanthropus profundus

that the difference in size between them comes from the fact that the tundra-dwellers were created large by the ancient genetic engineers as a precaution against heat loss in the cold north, and the islanders have become small over the past few thousand years as an evolutionary adaptation to their limited resources.

The islanders have no fear of the great creatures. They see them, as they see all living things that are not their own kind, as food. Nimbly they sprint down the beach towards them. Alerted by the crunching and rattling of the shingle under the tiny feet, the big tundra-dwellers see the little figures coming, and dimly perceive that there is some kind of danger. They turn to lope back into the forest, but they are too slow.

Two of them are caught by the legs and brought down with a crash. One is knocked senseless by the impact, the other is killed by quick bites to the neck and face. The killing is not easy. The hide is thick and covered with a woolly pelt, and there are deep layers of fat beneath.

It is the blood that the islanders want, and they gorge themselves on that of the slain tundra-dweller, balancing their feast with the carbohydrates from the fatty deposits. The corpse carries more food than the group of islanders can eat at one time, and having satisfied themselves they leave the remains to the white seabirds that have gathered on the shingle to watch the feast. This seems to the islanders to be a waste of food.

Together they pull the corpse of the second tundra-dweller up the shingle and into the shade of the forest, before it begins to decay in the sun or is eaten by scavengers. If only there were some way of keeping such a big creature alive while feeding from it. Then there would not be so much waste.

The massive form stirs; it is not dead at all, merely stunned. The islanders seize it by the limbs and pin it to the ground. They are not letting this one get away, nor are they going to let it die and rot before they need the food again.

SCHOOLS OF AQUATICS

In the green depths the school of aquatics works its way along the ocean bed. Spread out over a large area, each individual invisible to the next, the school keeps in tight contact by wails, clicks and twitterings – distinct but comprehensible sounds that form a language.

The body of creatures moves northwards, along the lines of magnetic force which are becoming more powerful again as the centuries go by. The direction they take is north, as geography goes, but the magnetic influence that they follow is towards the south. Since the time when the magnetic field disappeared, producing the fatal effects on the technological civilizations of the time, a great change has taken place deep within the globe. The magnetic field has re-established itself, but now there is a south pole where the north pole once was, and a north pole where the south pole once was. This reversal has little relevance to any of the creatures that now inhabit the world.

The water temperatures and currents are also changing, and this is leading to different patterns of fish movement around the globe. It may be that shoals of fish are gathering in areas unexplored by the aquatics, areas now free of pack-ice. If that proves to be the case, then it will make sense to move into those areas. The tropics are becoming over-fished.

The ocean never was particularly productive of food, considering that it covers more than two-thirds of the surface of the Earth. Back in the days of technological man, the living resources of the water were seized, exploited and lost in a short period of time. Since then nature has restocked, but the aquatics have always been there. Like the technological man that created them, the population of the aquatics has grown and grown. As they come to understand more about their own bodies, about diseases and injury and about reproduction, the birth-rate has exceeded the death-rate. Also, the life span of the individual has increased enormously and has been doing so for tens of thousands of years.

Around the coral reefs of the tropics the fish are vanishing, and the other valuable sea creatures are dying off. Undesirable and inedible species are moving in to replace them. The once beautiful and colourful fringing reefs, barrier reefs and atolls are now rapidly becoming dead skeletons of their former glory. It is not just the fault of the aquatics. The sea level is rising everywhere as well, and the tops of the reefs cannot grow quickly enough to keep pace with this. As the water becomes deeper and darker, the algae that grow with the corals and help them to feed are dying off, and the corals themselves are perishing. Although the aquatics cannot see colour (the rod cells in their eyes were developed at the expense of the cone cells in order to increase their low-light vision) they can see enough to know that their preferred environment is slowly dying. The aquatic colonies are everywhere in the shallows that surround the small tropical islands, becoming more and more crowded and more and more desperate for new resources, new food, new spaces.

That is why schools of them are moving northwards into the cooler waters; and others are turning their attention to a hostile environment – that above the surface of the ocean.

MELTING ICE

She will be able to remember her way home, she keeps telling herself. No matter how far the drifting mat of vegetation takes her or her family, she will remember her way back.

She, and the rest of her tribe, have been blessed in this. They have a knowledge that enables them to navigate to any place they want to go. The area where they live has been occupied by their ancestors since before the coming of the ice. Because of this they can actually remember the coming of the ice, and the places to which the different generations moved. It has all changed now that the ice is going back, leaving the landscape different from how it was before. Nevertheless they have always been able to travel to whatever place their ancestors knew would be good for food or shelter.

Now the ability had let them down. They wanted to go to a great river that their ancestors remembered from the dim past. Plenty of the fish were to be had in that river, and good shelters in the gorges through which it ran. However, when they arrived, the gorges had been gouged out by ice into a broad U-shaped valley with little shelter anywhere.

What is more, the river was in spate. The ice, away up at the head of the valley, must have been melting much more quickly than usual, and the water was hurtling down the valley floor in brown and white torrents, tearing at the river's bed and banks. The floor of the lower valley seemed to have been clear of ice for many years, because a coniferous forest had begun to grow in the soggy peaty soil. It was in this forest that the small group were resting when a sudden surge of the river wrenched away that part of the bank, trees and all. The intertwined roots and the solid trunks of the trees had bound the soil together and kept the whole chunk afloat as a kind of a raft, and the unfortunate group was carried away downstream.

Then night had fallen. The roaring of the river became quieter as it widened and slowed. There was no moon and the banks became invisible in the darkness.

She had panicked. With no visual landmarks her memory was not functioning. Another sense deep within

Water carries sound long distances, so the aquatics have been able to develop a complex system of communication. This keeps the school in contact when on the move, but allows sufficient space to feed.

50,000 YEARS HENCE

AQUATICS

Piscanthropus submarinus

As millennia pass, the aquatics become even more perfectly adapted to their seagoing existence. They become less bulky and more streamlined, with more efficient paddles and swimming organs. They begin to resemble the extinct seals and, like them, subsist on a diet of fish. However, they do not need to breath at the surface of the water. Their gills can extract all the oxygen they need from the sea. With the retreat of the pack-ice, aquatics move into unknown waters. This is essential if they are to survive a steady increase in population.

woodland-dweller
tundra-dweller
tropical tree-dweller
aquatics
boat people
cave-dweller
hunter symbiont
symbiont carrier
hibernator
seeker
social
islander

Homo sapiens sapiens
Homo aquaticus
Homo caelestis
Homo sapiens machinadumentum
Homo virgultis fabricatus
Homo glacis fabricatus
Homo silvis fabricatus
Homo campis fabricatus
Piscanthropus submarinus
Homo sapiens accessiomembrum
Homo menspropaeodorum
Speluncanthropus
Moderator bauli
Baulus moderatorum
Homo dormitor
Homo vates
Alvearanthropus desertus
Homo nanus
Nananthropus parasitus
Penarius pinguis
Piscator longidigitus
Formifossor angustus
Acudens ferox
Harenanthropus longipis
Giganthropus arbrofagus
Arbranthropus lentus
Piscanthropus profundus

her, a sense that should help her to find direction, was still working but it was very weak. She knew from experience that when she relied on this other sense and thought that a certain place was in one direction, it always turned out to be in the completely opposite direction. Something big must have changed completely since the days before the ice. She had had to resign herself to the possibility that she would never see her tribe again.

Now it is dawn, a cold grey dawn that brings nothing to warm the huddled and shivering figures on the floating island. The land has gone now and there is nothing to be seen but grey choppy sea. The drifting island consists of little more than a few trees and some trapped soil. There is no cover or shelter anywhere, let alone food.

The food will be irrelevant. They will all die of cold and exposure before they starve to death; unless they can remember something that their ancestors used to do under these circumstances.

There was something, she remembers vaguely.

It was something to do with rubbing sticks.

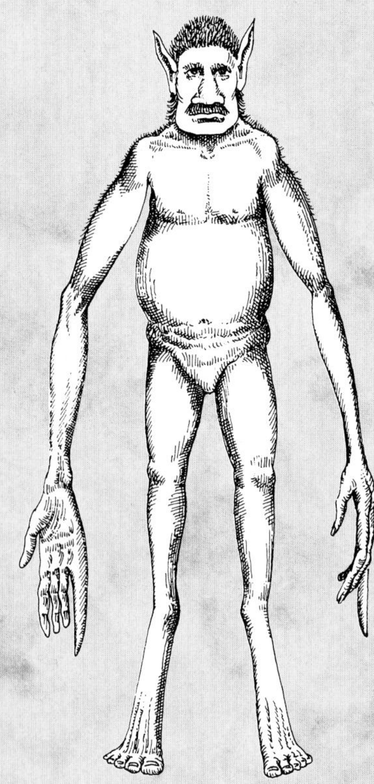

The socials evolved from the earlier plains-dwellers, the adult males are warriors and hunters.

500,000 YEARS HENCE

STRINGS OF SOCIALS

A string of figures winds rapidly through the arid scrub, kicking up clouds of dust from the red powdery soil. The sun is rising to the height of its heat, and soon the open semi-desert will be no place for any living thing. Despite their dark skins, and the protective covering of hair over their heads and backs, the socials would not be able to tolerate the shrivelling temperatures of midday. That is no problem, since at their speed the string will reach the Home before the conditions become too bad.

The spine of the string consists of about 30 youngsters, each carrying his or her allocated load of roots and tubers in woven bags. Moving parallel to them on both sides are about a dozen mature males, their sensitive eyes and ears scanning the red and grey landscape for potential enemies, their elbows bent and their huge bladed hands dangling in front of them ready for the defence of the string.

At the tail of the string two of the young gatherers are carrying a living creature between them. It is somewhat like one of the socials but smaller, and it does not have the long legs that allow the string to move so quickly. The two

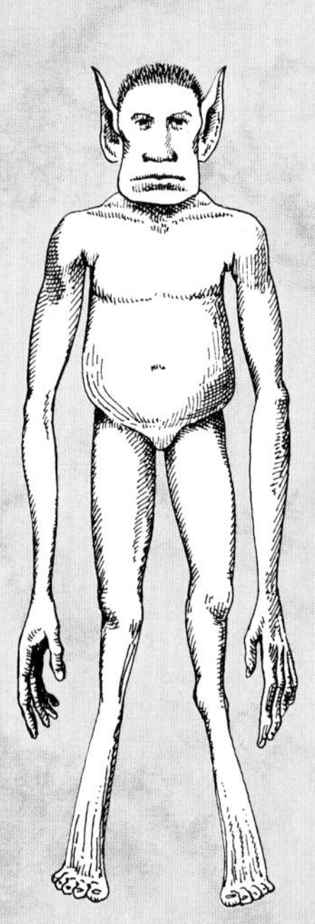

The juveniles of Alvearanthropus desertus *do most of the food-gathering.*

socials that carry it have interlocked their arms to form a kind of seat, and on this the creature perches with its arms around the necks of its supports. They treat this creature with care: it is their seeker.

Without a seeker the semi-desert would not yield up its tubers and roots, and its water deposits would remain hidden. Socials would use up their energy and time roaming the vast wastes in random attempts to find new food supplies. The seekers, although they are not part of the socials' family and lead their own lives within the Home, are a valuable part of the community.

The stringmaster pauses. There is something not quite right about the landscape ahead of them. He barks a single word and the whole string stops instinctively. They all drop down behind the scrubby bushes, to become invisible, but the cloud of their dust remains over their heads like a flag.

It is another gathering string, one from another community, encroaching on neighbouring gathering land.

With a few quietly grunted words, the stringmaster commands the young gatherers into a tight huddle, surrounded by about half of the fighting males, while the rest of the males spread out in a defensive arc facing the interlopers.

They need not have troubled with the stealth. The interlopers know they are there and are approaching in a determined advance, eschewing any cover. The stringmaster views the approach in dismay. This is no gathering string that has lost its way. It is a band of warrior males, without a juvenile gatherer or a seeker amongst them.

No further need for camouflage. The stringmaster barks orders that jab his own warriors into action. Up they leap from their cover and flail into the oncoming party. Instantly the stringmaster sees that his own fighters are outnumbered by about three to one, and so he calls forward those that are guarding the gatherers and their burdens. As for himself, he steps back out of the way of the fighting. He is too valuable to be wasted in the thick of the bloodshed.

They are still outnumbered but they fight on, kicking out with their elongated legs and feet, hacking downwards and sideways with the cutting blades of their hands, poking and gouging with their long fingers. The gristly hand-blades, originally designed to cut grass, can now shear through flesh and smash bone, and these are the main weapons of both sides. Severed limbs and heads lie in the dust, still pumping blood, as the defenders are forced back to the knot of helpless gatherers.

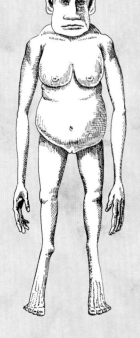

The females are confined to the community, looking after the young and the breeding mother.

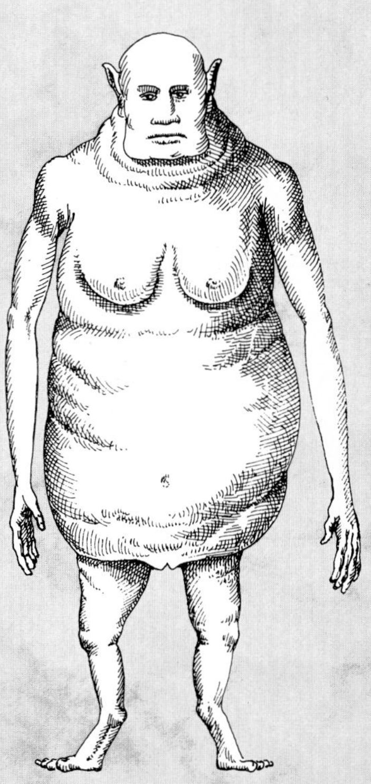

Only one female breeds at a time. The rest of the community revolves around her breeding cycle.

SOCIALS

Alvearanthropus desertus

Strictly-regulated and disciplined, social living produces a stable and efficient society essential for surviving in the more inhospitable places on the Earth's surface. However, genetic aberration occasionally produces individuals whose responses are not standard, and these introduce an element of chaos into the tightly-structured existence of such communities. Within the society, responses to danger are consistent and predictable – as are responses to any other stimuli. Functions are hierarchical and rigidly defined.

The hand-blades, originally developed to cut down thick grasses, have evolved as weapons making Alvearanthropus desertus *a dangerous foe. When socials fight, it is to defend territory.*

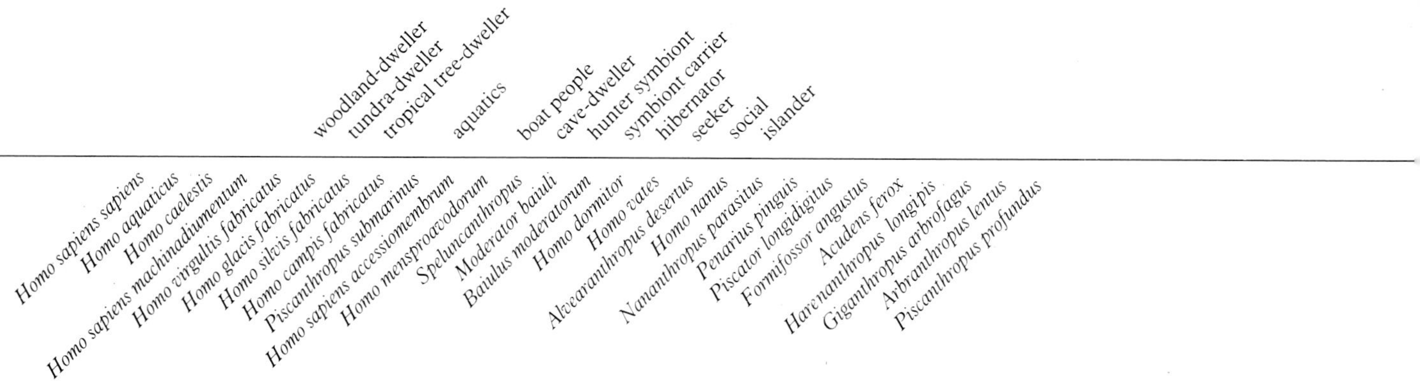

woodland-dweller
tundra-dweller
tropical tree-dweller
aquatics
boat people
cave-dweller
hunter symbiont
symbiont carrier
hibernator
seeker
social
islander

Homo sapiens sapiens
Homo aquaticus
Homo caelestis
Homo sapiens machinadiumentum
Homo virgultis fabricatus
Homo glacis fabricatus
Homo silvis fabricatus
Homo campis fabricatus
Piscanthropus submarinus
Homo sapiens accessiomembrum
Homo mensproavodorum
Speluncanthropus
Moderator baiuli
Baiulus moderatorum
Homo dormitor
Homo vates
Alocaranthropus desertus
Homo nanus
Nananthropus parasitus
Penarius pinguis
Piscator longidigitus
Formifossor angustus
Acudens ferox
Harenanthropus longipis
Giganthropus arbrofagus
Arbranthropus lentus
Piscanthropus profundus

The last warrior to fall is the stringmaster himself. He is happy to give his life for the defence of the string; less happy that it has been in vain and the string is lost. His last regret is that he will now never have the chance to mate with the mother.

After the defending warriors, the gatherers are easily slaughtered. Soon there is nothing left of the original string but the seeker, who stands unmoved by the carnage. The interloper's stringmaster addresses it and it agrees to lead them to its own Home. After all, it is a seeker. Seekers obey socials, whatever their Home.

The attacking stringmaster dispatches two of his warriors back to their own Home to summon young gatherers to take back the booty – warriors do not carry. He assigns about a third of his men to guard it where it lies. Then he organizes the remainder into a raiding string and has the seeker lead them towards the Home of their enemies. This string must move slowly, since the seeker cannot run as fast as the socials, and it cannot now be carried. Warriors do not carry.

Towards the blaze of noon, the bulk of the Home appears on the horizon. From a distance it would be unnoticeable. All that can be seen is a pair of ventilation chimneys that look just like the solid pointed towers of the beautiful and sacred insects that inhabit the entire region. The Home itself is in a hollow, an impenetrable fortress. Smooth walls, with no hand- or footholds, red and hard as bone, curve upwards enclosing the entire colony in an impregnable dome, the shape of a tuber. Only the two tall chimneys at the top break the symmetry. Near the top a crack in the structure is being repaired by a small group of gatherers, moist red clay being kneaded and pressed into the damaged area. In this vast structure are the mother, the infants, the juvenile gatherers, the female nurses, an unknown number of male warriors, old male drones, a ghetto of seekers and, most important for the raiders, the foodstores that would sustain them all.

The raiding stringmaster, having hidden his warriors and crept as close as he dares, peers over the rise of the land not far from the Home. It is as well guarded as his own. Each of the ground-level entrances is guarded by several warriors, and most likely many more warriors are housed in chambers close to the entrances. Breaking in is going to be difficult.

The vague stirrings of an idea occur to him. He often has ideas. Even when he was a mere gatherer he did, but it was difficult to work on them then, when everything that he did was prescribed, regimented and expected of him. Likewise, when he had grown to a warrior and his female siblings had gone to be nurses those ideas came to him. Only in the heat of battle, when an individual could act on his own initiative for the good of the Home, could any of them come to fruition. Most of the time events showed that his ideas had been justified. That is why he is now a stringmaster. This idea, however, is something quite novel – disturbingly so.

Stealthily he makes his way back to his warriors and the captured seeker. With much difficulty, through the few words they possess, he gives the seeker his instructions. The seeker is puzzled. It takes a long time to convince him of what is required, as this is something new to him as well. Eventually he seems to understand and goes off towards the Home.

The guardian warriors at one of the entrances start into attentiveness as they see the lone seeker scrambling down the dusty slope towards them. They demand to know what he is doing. Dutifully the seeker states that the string is under attack, not far away, in the direction from which he has come. When he is asked for more details, however, he is blocked. He was not told to report anything more. As these warriors ask him questions he becomes more and more confused. The answers he should give are in conflict with the statement he was told to make. He was given orders by socials. Now he is asked questions by socials that would confuse the first orders. He throws his arms over his head and collapses to the ground. He cannot understand what is happening.

Nor can the defending warriors. What they have understood is the report that one of their strings is under attack. They rouse the other warriors of the Home and form themselves into a fighting string, running out in the direction indicated by the gibbering seeker.

Once they have gone and things are quiet, the raiding stringmaster brings his warriors stealthily from the other direction to the abandoned entrance. He picks up the cowering seeker and shakes him back into attention. Then, preceded by the seeker, the raiding party enters the Home.

There are still warriors in the chamber behind the entrance, but these are soon silenced, and the raiders make their way into the interior. Pushing the unhappy seeker before him, the stringmaster and his warriors penetrate deeper and deeper into the Home. The air becomes heavier and stuffier. This is to be expected. As the females grow to be nurses and, in a few instances, mothers, they spend their time deep in the airless tunnels and chambers. Their metabolism slows, allowing them to consume less air and less food, and devote their lives to feeding mother and infants.

The seeker dodges out of the passage and into a side chamber, illuminated by a dusty shaft of light slanting through a hole in the outside wall. A great commotion arises. This is part of the seekers' own quarters, a rambling disorganized muddle of chambers and passages within the walls of the Home, a place of chaos and random life where these low creatures mate and play at will, fed and cleaned constantly by the Home's nurses. The seekers, despite their disgusting habits and lifestyles, are essential to the life of the Home.

The dark bobbing shapes of his companions welcome him back but are then thrown into consternation by the appearance of strange warriors behind him. A nurse, bringing the seekers their daily ration of food, is shocked into immobility and stares stupidly at the raiders. A bowl of chewed roots and flattened insects falls from her long hands. They kill her immediately but leave the seekers alone. The captured seeker has now collapsed in terror and confusion amongst his companions and will obviously be of no further use. The stringmaster and his men push onwards and downwards, feeling their way in the darkness now. Occasionally they come across the soft slow body of a nurse, or the active one of a juvenile, and these they kill without hesitation. Those that are nimble enough to escape are ignored. The raiders are after more important prey.

Eventually, in the dimly-lit chamber beneath one of the ventilation chimneys, they find her: enormous and reclining, fat with obesity and pregnancy, her hairless skin over folds of fat glistening dimly in the gloom – the mother.

Around her move a dozen pale nurses carrying in food and taking away waste. Slow drones, their weapon hands hanging long unused by their sides, stare stupidly at the intrusion. All cluster around the mother in a vain attempt at protection.

The raiders move in. The nurses put up no fight at all, but the drones, remembering their glorious days as warriors, make a token struggle – and perish. At last the prize is won. In the dimness the mother pathetically tries to pull her great bulk away, on her stunted legs and wizened arms. She lets out a plaintive wail as the raiders fall upon her, and she dies under their hacking hands.

Not long afterwards, the mother's body hangs head-down from the partly-repaired crack on the outer wall of the Home. The stringmaster stands in triumph above it. All the fighting is done now. The returning strings of defending warriors, those that had been lured from the Home by false information, are totally demoralized by the sight. Their tightly co-ordinated groups break up and scatter, and the individuals wander off into the arid landscape, inevitably to die.

The Home is the stringmaster's now. Normally he would send messengers to their own Home, and they would return with gatherers who would strip the captured place bare and carry all the food and the seekers back to their own, thus expanding their hunting territory.

This time, though, he is going to do something different. This whole incident has been different so far. There has never been a Home won over by using deceit, a totally alien concept amongst the socials. Their language is simple, but it has always allowed for individuals to express themselves, for stringmasters to communicate orders to warriors and seekers, and for gatherers to describe the whereabouts of food supplies and their dimensions. This is the first time that their language has been used in a deliberate way to deceive. It is indeed a new and useful development, showing great promise for the future.

The other difference in this campaign is that this Home is not going to be destroyed. There will still be young nurses cowering in the tunnels and warrens below, one of which he will make the new mother. The other nurses and the few juvenile gatherers that are left will naturally be loyal to her, and his warriors will remain loyal to him, or he hopes that they will until he can raise new ones of his own. He will send deceitful word to his former Home that his own string has been wiped out, so he will not be missed.

For the first time a new Home will be established, not by a mating pair cast out of a single Home, but by uniting two strong Homes, drawing on the strengths of each.

BOATBUILDERS

The working of metals had been a forgotten art; but then it was remembered – and forbidden. The making of boats had likewise been forgotten and then remembered and had likewise been forbidden.

Now those who have dared to practise these skills are dispossessed. The boats they made carry them to safety, away from the anger of the remainder of their people.

The boats are sturdily-built, of planks cut by metal tools and pinned together with wooden pegs. Someday they will

500,000 YEARS HENCE

BOAT PEOPLE

Homo mensproavodorum

The inherited skills that began with the making of fire threw up the memory of boatbuilding. With the forbidden memory came an instinctive drive to use it. Descendents of the memory people, the boatbuilders can now travel freely to colonize habitats not their own. Sharp teeth and hooked claws are their natural weapons, but with the discovery of metal comes the blade.

The aquatics have devised a method of returning briefly to the land, carrying their own saltwater environment within a tough sphere of gel. Faced with enemies, they are slow and vulnerable.

be able to build them of metal – if this is permitted. For now the five boats are carrying the 43 individuals who represent the only group of beings in the world with the courage to use the remembered knowledge of their ancestors. The woven sails bulge with the wind that they know will carry them to the islands in the warmer regions of the globe.

It is not that they lack the conscience and moral terror of the rest of their people, just that they feel strong enough to overcome any danger. They know, deep inside them, that the knowledge their ancestors gained, generation by generation, eventually destroyed them. They know that their ancestors made things, that they took power from the sun and the sea, from the ancient concentrated remains of life, from the breakdown of the very forces that held matter together. With this power they took metals, food and other materials from the solid Earth and from the living creatures that existed on it. They were able to increase their life spans, eradicate the diseases and accidents that held populations in check, and spread over the whole surface of the Earth. Eventually the Earth had become too crowded and burdened to carry them, and they perished under the weight of their own technical cleverness. All this they remembered, although they hardly understood it; but the inherited memory of the loss of everything that their ancestors had achieved was enough to forbid the use of the inherited memory of the means of achieving it.

All abided by this, except for the boatbuilders, who continually flouted their people's taboo on using their ancestors' knowledge, and were persecuted for it. They fought back with blades, but the overwhelming hostility had driven them away from their fertile homeland. Now they are on the run, but it may not be for long. Many of their boats have been left behind, and it seems likely that the more zealous of their enemies will come in pursuit. Although boatbuilding is forbidden, sailing them may not be – and everybody shares the memory of how to sail.

What is more, their choice of destination has been made on the basis of inherited memory. Their pursuers, using the same mix of ideas, inspiration and basic knowledge that the inherited memory entails, will come to the same conclusions. There is no such thing as secrecy now.

After many days of steady winds, the fugitives see the first of the islands. It is as they expect. The first sign is a cloud on the horizon; blue hills appear next, then the green of lowland vegetation, and finally the white streak of beach. All is as predicted – except for the bubbles.

Several shining bouncing globes are moving up the beach. The puzzlement that they produce in the boatbuilders is short-lived, however, as the boats are caught in the rising swell of the shallowing sea. The waves that have pulsed unnoticed across the open ocean are now funnelled and magnified as the seabed shallows, building up into steep walls of green water that curl over and crash into an oblivion of sparkling white spray and surge, hissing up the hot sandy beach. In this turmoil the boats heave upwards, dive into the hollows and are flung towards the land. As the prows crunch into the beach, the boatbuilders jump out, splashing ankle-deep in foam and sand, and drag their vessels to safety. Then, when all are safely ashore, they collapse onto the beach in joy and exhaustion. Although the voyage was completely predictable, because of their common memory, they have been very uneasy during their days at sea. That was not their environment at all.

One of their females notices it first: the huge translucent sphere beneath a sagging palm tree at the head of the beach. They had all seen the bubbles from the sea, but had ignored and then forgotten them. It was always the way that the inherited memory was more powerful than that developed by the individual. In size, the sphere could probably be encompassed by the outstretched arms of three people. It is shiny with a greenish tinge, and its base is spread and flattened by its own weight. Its outer covering seems flexible and the whole thing wobbles as it rolls slowly down the beach towards them. Sand adheres to its outside as it moves, but dries and drops away very quickly.

The female who first saw it stands and watches it roll right up to her. All watch, to see what happens next. Inherited memory cannot guide them now. Before there is time for reaction, a silvery arm shoots out of the side of the sphere, seizes her hand and tugs it inside. Then it starts rolling towards the water's edge, dragging the surprised female with it. When she realizes what is happening she begins to scream, but she and the sphere disappear beneath the surf before anyone can do anything about it.

The travellers stare after her, stupidly. Then several more of the spheres appear at the head of the beach. They do not seem intent on attack – they roll towards the sea, avoiding the party. Anger, an emotion not often felt by the boatbuilders, surges to the surface, like one of the bursting waves, and as one they launch themselves in a revenge attack at the nearest sphere. Surrounded, the sphere cannot move, but it seems to waver, this way and that, to try to break free. Its surface is yielding but too tough to be penetrable. Blows and punches are absorbed and bounce right

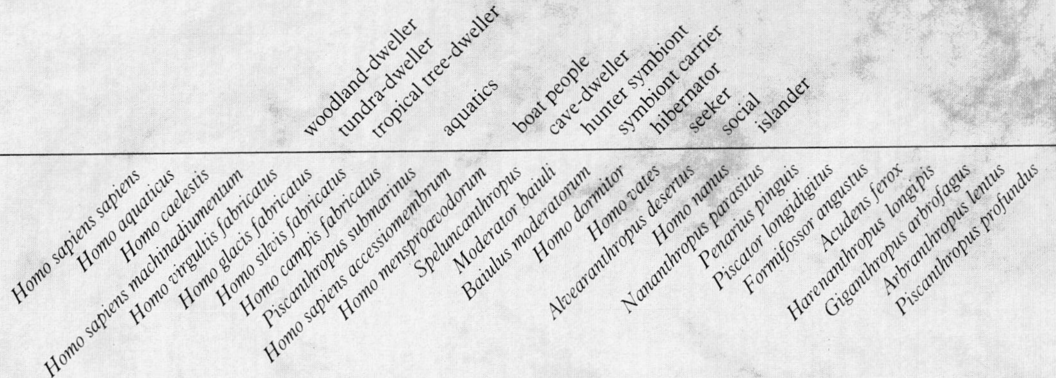

woodland-dweller
tundra-dweller
tropical tree-dweller
aquatics
boat people
cave-dweller
hunter symbiont
symbiont carrier
hibernator
seeker
social
islander

Homo sapiens sapiens
Homo aquaticus
Homo caelestis
Homo sapiens machinadiumentum
Homo virgultis fabricatus
Homo glacis fabricatus
Homo silvis fabricatus
Homo campis fabricatus
Piscanthropus submarinus
Homo sapiens accessiomembrum
Homo mensproavodorum
Speluncanthropus
Moderator bauli
Baiulus moderatorum
Homo dormitor
Homo vates
Alvearanthropus desertus
Homo nanus
Nananthropus parasitus
Penarius pinguis
Piscator longidigitus
Fornifossor angustus
Acudens ferox
Harenanthropus longipis
Giganthropus arbrofagus
Arbranthropus lentus
Piscanthropus profundus

back. Then one of the boatbuilders brings a blade from the boat and plunges it into the glistening surface.

The sphere bursts, and a rush of salty water gushes over the attackers and sinks into the dry sand. The punctured surface has collapsed into slimy gel, releasing seawater. In the middle of the stain lies a strange creature, gasping.

Like them it has a black skin, but the skin is completely smooth and hairless. The head is like that of a fish, with big eyes that do not seem to be functioning in air. The mouth is huge and gaping. No neck separates the bulbous head from the streamlined body. Gills on the chest flap ineffectively, and the body narrows to a paddled tail. It is the arms, however, that are most remarkable: they are human arms, complete with hands and fingers. The thing flaps about on the beach pathetically as it slowly dies of suffocation.

The sea creature has devised some means of coming onto land and bringing its own environment with it. If these islands are now the domain of these creatures it is going to be difficult to settle here, for they have proved to be undeniably hostile.

Moreover, what will happen when the boatbuilders' pursuers arrive?

1 MILLION YEARS HENCE

HUNTERS AND CARRIERS

The leader starts from his sleep because his carrier is uttering grunts of alarm. Dawn is almost here, and the sun is already shining on the highest snow-clad peaks of the range, although the valleys are still in deep purple shadow. The mountain birds have set up their calling and the short grass beneath him is damp with dew, but his fine pelt keeps him from the chill.

What has disturbed his carrier?

He unwinds his long limbs from around his female and rises to his spindly legs in the cold half-light. Most of the rest of his clan, hunters and carriers, are asleep. He can see the hunters, huddling in pairs or with their children on the slope. The huge white shaggy forms of the carriers are more visible, forming a loose defensive circle around the group. His own carrier, whom he thinks of as Oyo, is awake and alert, disturbed by something that he cannot see.

Could it be one of the distant creatures from the far lowlands? It is not really likely, since they rarely come this far up the mountains, particularly at this time of year. Nor is it likely to be one of the big birds. They do not attack so early in the morning.

On all fours (his usual posture) the leader trots around the group to check that all is well, and realizes that he is not the only one awake. At the far side of the circle two hunters are mating, with gentle noises. He looks around and, true enough, their carriers are mating too – a rougher exercise accompanied by hoarse grunts. Certainly nothing is amiss here.

He scrambles over to where his carrier stands, a white silent dutiful column. Without a word he scrambles up the fur and onto its back, resting his narrow chin upon its usual spot on the broad cranium. The massive shaggy arms come up and clutch him firmly. Now he can communicate by thought, without clumsy language.

With thought he commands the great Oyo to turn slowly so that he can scan the lightening landscape. He does not realize it, but this is far from the landscape of his ancestors. The hunters and the tundra-dwellers first came together on the chill wastelands bordering the retreating northern icecap. The tundra-dwellers were well adapted to the cold, and their great bodies could generate enough heat to keep the slim-limbed hunters warm. The hunters for their part were nimble enough to catch the most evasive of food, and to catch enough to feed both of them. Together they made up more than the sum of each. There is no icecap left now and no tundra, and nowhere in the lowlands are there any environments that suit them; the forests and woodlands are more suited to different humanoid creatures altogether. Only in a few places, on chilly peaks and in cool mountain valleys, are conditions still right. In these isolated places the symbionts linger, marooned as the colder conditions withdrew up the mountains, and towards the pole where they disappeared.

Nevertheless there is still a good living in the mountains: plenty of small mammals and birds for the hunters to hunt for themselves and to share with their carriers, and plenty of grasses, mosses and lichens for the carriers to scrape up and share with their hunters. Hunters and carriers mate at the same time – the mating of a pair of hunters inducing mating in their respective carriers, and vice versa. This usually results in the birth of a hunter baby at the same time as a carrier baby. Both babies are carried by the parent carriers for about six years, at the end of which the young Hunters choose their own carrier of the same age and same sex.

The family groups move with the seasons, from the grassy slopes of the valleys in the winter and spring to the flower-strewn bluffs and crags of the peaks in the summer and autumn. The habitable areas, although productive, are few and scattered, and the tribes of symbionts have their own ranges.

Down the brightening slope, with the grey mist of the valley behind it, stands a stranger. That is what had disturbed Oyo: the massive shape of a carrier, with the squat hummock of a hunter lying over its shoulders and head.

With a burst of thought the leader asks Oyo if it recognizes the newcomer, but the dim-witted reply is inconclusive (direct questions like this between hunter and carrier rarely yield anything useful). The stranger strides purposefully up the hill towards them.

It is a challenge. Evidently this is a rogue male, thrown out of a clan, possibly even thrown out of the leader's own clan some time in the past. Wherever it came from its intentions are now clear. With thin yells and reedy shouts – strange noises to be coming from the huge bulk of a symbiont – the newcomer utters its threats and challenges. The leader replies in like voice.

The result is ritual. The hunters pull themselves back from the great heads of their carriers, and hang on tightly with their own hands to the long fur of the shoulders and back. This frees the carriers' arms for the combat. Then, spurred on by the hunter's thoughts, the great carriers that bear them wade into each other, striking, slapping and pushing with the flats of their vast hands.

Most of the blows land harmlessly on the great areas of muscle and fur on the chest and forearms. An occasional blow that lands on the face brings blood from the nose or the lip, but does no serious damage. This sparring will continue until one of the combatants, usually the attacker, tires and turns away, or else falls over separating hunter from carrier.

On this occasion the combat is quite predictable. Although the attacker's carrier is big (bigger than Oyo, in fact) the hunter does not have the mental skill to guide its blows and punches to best effect. If, by chance, he did become the leader of the clan, its future would not look good. Mental skill is needed by a leader in order to judge the timing of the fruiting of food plants, and to plan the routes of migration.

This time, however, the leader's mental agility is not proving to be enough to counteract the strength of the attack. Oyo is cringing in pain from the bruises and cuts from the blows which the opponent's carrier is hammering down with particular ferocity. The sharpness of the pain and fear is picked up by the leader, through the same nerves and ganglions along which he gives Oyo his orders.

It is no good! He is going to have to step down. Oyo will die if this continues, so after all these years he must relinquish the leadership of the clan. He had not anticipated anything like this. Yesterday he was at the height of his powers and virility; now he must give way to a younger symbiont. He will live out his days as an old and revered clan member, nothing else.

He steps back and turns round, presenting his naked back to his opponent: the age-old sign of surrender. The clan now belongs to the attacker.

What happens then is totally unexpected, and quite against any tradition. The opponent's carrier seizes him by the exposed neck and shoulder with its huge hands. Strange thoughts and emotions burst through him from the contact with his enemy – thoughts of rage and hate and uncontrolled violence from the carrier, unchecked by the feeble commands of the controlling hunter.

The leader is torn free from Oyo's fur and flung onto the ground. The flood of alien thoughts ceases, as do the sensations of pain and panic from Oyo. It is just as well. The attacking carrier brings down its great hands on Oyo's back and shoulders, flinging the dear creature to the ground, and wrenches back its head, breaking its neck.

The silence that follows is not just the silence of the horrified clan, who have been roused from their sleep and are watching the fight earnestly. Nor is it the silence of the hillside, produced when the birds are quietened by the violence of alarming events. It is the aching silence of loneliness.

Oyo is gone. Half of the leader's being is dead, and the other half must follow soon. He can no longer be a part of the clan, but must seek out a life of his own and exist as best he can.

This is always a failure. A hunter without a carrier, like a carrier without a hunter, is always dead within a few days.

Yet cutting through the searing grief is an even more troubling thought. The clan – his clan – is now in the charge of a symbiont that consists of a powerful violent carrier that cannot be controlled by its hunter. The hunter, as well as being weak, does not have the mental versatility to lead a clan. That much was evident during the fight. It is not just his own death and that of Oyo he mourns, but the death of his entire clan and family.

As the icecaps retreated, the symbiont tundra-dwellers – Baiulus moderatorum *– retreated, living at high altitudes and near the poles. Nowhere else is there a suitable environment.*

ONE MILLION YEARS HENCE

HUNTER SYMBIONT

Moderator baiuli

Communication between hunter and carrier has been simplified to a telepathic link – the huge slow-moving tundra-dwellers controlled directly by the weaker but agile-minded hunters. Fights, when they happen, are usually ritual. Death is unexpected.

AQUATICS

Piscanthropus submarinus

As the aquatics spend more time on land, their tough protective bubbles refine and become more efficient. Eventually the gel becomes form-fitting, holding the thinnest layer of life-giving seawater against the aquatic's body. This covering is enough to keep the skin moist, and to absorb oxygen from the air which is then absorbed through the gills. A steady increase in population among the aquatics has led to food shortages and famine. With the sea stripped bare, the aquatics face a hostile environment.

With food in short supply competition between species becomes, literally, a matter of life and death. Once out of the water, aquatics labour under their own weight.

AQUATIC HARVESTERS

There is no more food growing here; it has all been cleared out. The ravaged soil has scraggy shoots sticking out of it, but it will be a long time before these grow and bear anything worth eating. Dead tree trunks stand gaunt and stripped, harsh splintery wood, killed by greed – no, not by greed, by necessity. The leaves had to be taken to feed the aquatics, but now the trek from the sea to the food is becoming longer and longer.

Ghloob peers through the watery film and the gelatinous envelope over his eyes. This work is dangerous and unpleasant, but the days of easy and pleasant life disappeared long before his birth. It is said that once the sea, their home, supplied all their needs, but then their numbers became too many, and all the food was gone. Famine raged. Whole populations perished and sank into the dark deeps. Sometimes after famine, the fish, krill and plankton would return, but this food source was never enough. As soon as it came back it was exploited and destroyed once more. Nothing can be done about it: if they want to survive they have to eat; if they eat they lose what they have and die.

It is as if there can never be a balance. They live there but they intrude on the natural system of things; and nothing that they do will make it any better.

Now they are exploiting the land as well, thanks to the algal mats that they have developed. Filamentous algae forming a fine mesh, impervious to water but permeable to air, can be induced to make shapes that will hold water. An aquatic can ascend from the ocean into the harsh sunlight and thin air above, still immersed in seawater, but contained in a flexible gelatinous envelope of algae filaments. Air passing through the envelope keeps the water aerated, and the aquatic neither desiccates nor suffocates, as long as the envelope holds.

Progress has been considerable. When the technique was first developed the envelope had to be spherical, holding a vast quantity of water. The adventurous aquatic moved along in this, rolling the squashy sphere around him, a cumbersome process. Now, and Ghloob cannot remember when it was otherwise, the envelope is form-fitting. Only the thinnest of water layers surrounds him and protects him from the harsh world of the outside. Movement is still difficult, though, and always will be. He feels his own weight – an unknown sensation in his natural home – and he must pull his elongated body along the ground with his arms. If he is carrying something, he must

The flexible envelope is made of gelantinous algae filaments and filled with seawater. Its close fit allows more freedom of movement than the earlier bubble.

wriggle along as best he can. Then he has to take care that the jagged denuded ground does not rip the envelope. No, this is not natural.

It has been good enough, though, to allow the aquatics to exploit all the lands that border the ocean. They sweep them clean of any growing or living thing, and do not give anything time to regrow. The teeming populations below the waves cannot wait.

In the distance, glimpsed hazily through the algal membrane, loom shapes that could be trees, or they could just as well be naked rocks. Aquatics had no colour vision built into them when they were engineered, and none has evolved since.

He cannot communicate with his companions, but he hopes that his actions will be clear. He humps his long body, in its glistening envelope, in the direction of the shapes. The three others that are like him turn and follow. The fourth, the one encased in the spherical bubble that looks like one of the originals, is guided along by them. It is he who will enfold and carry home any food that they find.

They are travelling up a slope, which is not good. Distance from the sea is one thing, but height above its surface is another matter altogether. The aquatics live happily with the pressures experienced in the top layers of the ocean, but they are under considerable strain when exposed to the reduced pressures above the surface. To go any higher would produce all sorts of unwelcome effects in their tissues. An abrupt contour line, above which vegetation grows freely in many parts of the world, marks the limit of aquatic exploitation.

Beyond this contour line live the land people – strange beings who neither understand nor tolerate the aquatics.

There are the tree-dwellers, of whom the aquatics know little. They keep themselves in the branches away above. Aquatics rarely look upwards (it is difficult for them to do so), and so these beings are rarely seen.

Then there are the ground-dwellers. Savage and hostile, they feed in the undergrowth and the long-growing vegetation – the very materials that the aquatics harvest. Gangs of them have been known to burst out of hiding and set themselves upon harvesting groups, tearing at their protective membranes with claws and teeth, and sometimes inflicting some damage.

There is also the massive compound being, a huge basic creature, bloated and misshapen, lumbering through the forests with four or five spindly little figures attached to it, embedded in it, seeming to live off its flesh. These beings cause no trouble; in fact, they sometimes blunder out into harvesting parties where they are particularly vulnerable. In the open they are easily brought down and the moving reef of flesh can be killed by blows from an agile aquatic or drowned by being dragged within a membrane. The small attached creatures – tiny wizened bodies with spindly crablike legs and enormous mouths – become strangely pathetic without their mount and scuttle clumsily for cover. There is good eating on the fat creature and it is always borne back to the sea as a prize.

Finally there are the fighters, which are a menace, because they seem quite at home on the devastated areas left behind after harvesting. Their home is in the drier parts of the landmasses, where little grows anyway. They are organized, and many dozen can attack at once, moving as a single entity as if controlled by a single mind. Their forelimbs are cruel cutting weapons that can slice through a living membrane with a blow and kill the aquatic inside, so this time it is the aquatics who are the prey and their wet dead bodies are dragged away to the fighters' citadels. Of late, the attacks have been so organized that it is evident that the skirmishes are no longer defensive. Parties sally out with the firm intention of capturing and killing the harvesting aquatics. These beings must be left alone, and their domains avoided at any cost.

The shapes prove to be trees after all, but the undergrowth beneath them is patchy, curled and dead. Since the area down to the ocean has been devastated and left open to the sky, the air moving off the sea has swept in through the branches and between the trunks, drying up and battering the fragile stems and shrivelling up the leaves. Loose sand and dust from the bare lands has gusted in, suffocating the more delicate types. There is little to be harvested here, but what there is must be taken.

Ghloob and his companions reach out their hands through the membranes and snatch up whatever is growing. Anything that is organic, and contains proteins and carbohydrates, can be used as the basis for food, however tough, however unpalatable. Bundles of leaves, stems, sticks, insects, slugs – anything – are caught up and passed into the sphere of the gathering aquatic. Small punctures in the membranes, like those caused when hands pass through, seal up immediately and there is little or no moisture loss.

Before long the cache within the spherical bubble has become quite large; large enough to take back. The five of them turn to make their laborious way back to their ocean

Homo sapiens sapiens
Homo aquaticus
Homo caelestis
Homo sapiens machinadiumentum
Homo virgultis fabricatus
Homo glacis fabricatus
Homo silvis fabricatus
Homo campis fabricatus
Piscanthropus submarinus
Homo sapiens accessiomembrum
Homo mensproavodorum
Speluncanthropus
Moderator baiuli
Baiulus moderatorum
Homo dormitor
Homo vates
Alvearanthropus desertus
Homo nanus
Nananthropus parasitus
Penarius pinguis
Piscator longidigitus
Formifossor angustus
Acudens ferox
Harenanthropus longipis
Giganthropus arbrofagus
Arbranthropus lentus
Piscanthropus profundus

home, glistening welcomingly away on the horizon.

No sooner have they left the shade of the dying trees, and begun their long slow descent, than Ghloob sees something at the periphery of his vision, something moving.

Slowly he turns his head. Ground-dwellers! A whole pack of them! They are running towards the aquatics, waving sticks of some kind. His companions see the danger at the same time, and try to move more quickly. However, their laborious humping motion is not conducive to haste, and anyway they cannot move faster than the spherical bubble containing their harvest – the only reason they are here in this hostile environment in the first place. The ground-dwellers quickly surround them, and as their hazy shapes appear before him Ghloob notices something different about them. They are each carrying something: something like a blade at the end of a stick.

Ghloob has not much time to notice anything else, as he ducks out of the way to avoid them, but after heaving himself along the ground for some distance he turns to look back. The ground-dwellers have all set upon one of his companions. They have plunged their weapons into his membrane and are pulling it apart. With two creatures pulling in different directions this turns out to be very easy, and the membrane collapses in a gush of water leaving the stranded aquatic gasping in the circle of wet mud.

Ghloob and the others crawl frantically away, towards the tempting but distant sea, panic rising within them; with good reason, for the party of ground-dwellers leave the dying aquatic and come running after the straggler of the group and fling themselves upon him. Ghloob does not stay to watch this time, but keeps wriggling.

With every jump and jerk he expects to be attacked from behind, and his membrane torn away from him. The waves of the ocean come closer and closer, but agonizingly slowly. Will he make it before they catch him? He tries not to think about it, and keeps going.

With an intense feeling of joy he feels the pressure of the first wave close around him. He is safe, and at last he can look around. The bubble with one of his companions and the gathered food has reached the sea. The food is also safe, but at what cost? Three companions are lost – punctured, dehydrated and slaughtered on the distant dusty dryness.

The ground-dwellers have never fought like this before. Perhaps the aquatic harvesting has had such an effect on their lifestyle that they have had to adopt these extreme measures to fight back. Maybe the conflict and strife have forced them to find new ways of living and organizing themselves just to survive.

Ghloob's algal envelope dissipates now that he is fully submerged, and with graceful movements he descends the sloping seabed until he is below the push and pull of the waves, and home. Now he has time to ponder. Is this organization and use of weapons by the ground-dwellers to be a feature of all such attacks in the future? Has the aquatics' exploitation of the land made even that more hazardous? Is there nothing that they can do to feed their people without making things worse and worse and worse, and destroying everything that they have? Is this to be the continuing fate of intelligent life above and below the water?

2 MILLION YEARS HENCE

TRAVELLERS

The food will be there, and can be taken, as the travellers know. Every year the enclosures ripen, the planters awake, feed, repair the enclosures if necessary, plant the new seed and return to their slumbers once more. The secret is for the travellers to time the journey so as to arrive before the planters rouse from their long sleep. The planters are supposed to be a very ancient race, and each one lives for many hundreds of years – if 'live' is the right word. How can you be living if more than nine-tenths of your time is spent asleep?

How did this come about? It probably goes back to the time when the differences between the cold times and the warm times were much greater than they are now. There have always been animals that have hibernated – slowed down their systems and gone to sleep during the coldest time of the year. These creatures usually gather their food and store it, waking up and eating from time to time; or else they eat so much when they are awake that they build up stores of fat that nourish them while they sleep. The planters were once normal, like the travellers, but probably not so intelligent. Back when the ice had just shrivelled up from the continents and the 'winters' were still cold, they developed the ability to sleep away the harshest of conditions, and they stored up food as well. Some of the seeds and grains that they stored would have germinated by the time the stores were opened; if the hibernation time were long enough they may even have fruited again. As the cen-

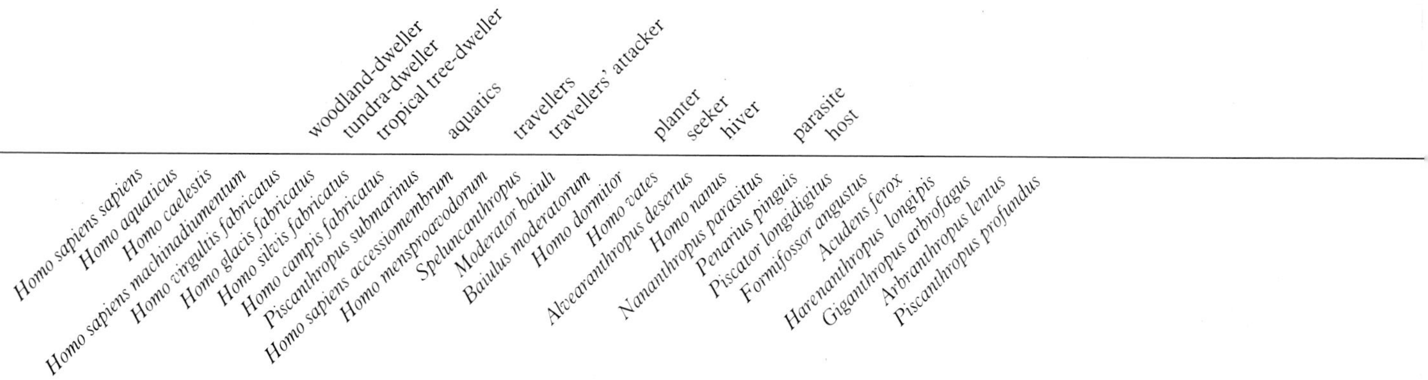

woodland-dweller
tundra-dweller
tropical tree-dweller
aquatics
travellers
travellers' attacker
planter
seeker
hiver
parasite
host

Homo sapiens sapiens
Homo aquaticus
Homo caelestis
Homo sapiens machinadiumentum
Homo virgultis fabricatus
Homo glacis fabricatus
Homo silvis fabricatus
Homo campis fabricatus
Piscanthropus submarinus
Homo sapiens accessiomembrum
Homo mensproacodorum
Speluncanthropus
Moderator baiuli
Baiulus moderatorum
Homo dormitor
Homo vates
Alveanthropus desertus
Homo nanus
Nananthropus parasitus
Penarius pinguis
Piscator longidigitus
Formifossor angustus
Acudens ferox
Harenanthropus longipis
Giganthropus arbrofagus
Arbranthropus lentus
Piscanthropus profundus

turies and millennia passed, the planters developed the ability to remain suspended until harvest time, when they would come out and eat, plant the next crop and retire again.

The travellers knew that it was possible for such things to happen. Vaguely they remembered the knowledge that their ancestors had possessed, knowledge about changing conditions and changing life.

There must never be any dealings with the planters. The planters build their enclosures, and use the growing vegetation not just for food. They gather their food from where it grows, but also plant it in places that will be more convenient for them to collect it from. They build walls and roofs of stone and wood to protect what they have done, just as their remote ancestors did. It was the beginning of the changes that eventually destroyed everything – the land, the living things, themselves. Now nothing must be altered, nothing must be built, nothing must be changed from its natural state; that is the credo of the travellers.

It is a sign of their strength that they know how to make their life easier, but ignore the knowledge. Any one of them has enough inherited knowledge to dig the burning stones or the naturally-distilled organic fluid from the ground (if indeed there are any deposits of these left) and use their heat to melt down the metal minerals. They could all break down the substances from the rocks and use them for many varied purposes. They know that it is possible to fly to the moon and stars, and they know how to do it; but they will not. They will not call down the destruction once more.

It is not just their memories that impress this credo upon them. Wherever they travel, through the lush forests and woodlands or across the open plains and deserts, they see the dismal results. In a forested valley, where they remember once stood a city, the rocks that outcrop in the slopes of the stream gullies are not natural. They are man-made, sometimes with unnatural angles and faces that have miraculously survived 2 million years of burial. The soil here is stained and streaked with red and green where the vast volumes of metal that went into the artefacts have oxidized away to dust. The area is disgustingly unnatural, and avoided.

Elsewhere lie similar remains that are lethal to any creature that passes close by. Even now, 2 million years later, the technological overproduction of their ancestors has the power to kill. Nothing appears at the surface here, but not far down lies the disintegrated ruin of some vast structure. So great have been the natural forces of erosion and decay that nothing recognizable of the original structure remains even underground, but some of the raw materials still lie there, emitting a deadly force. Anyone crossing this area sickens and dies. The travellers remember that it was something to do with the generation of energy.

This is why the travellers despise the dark-minded creatures, their distant relatives, with whom they share the planet but who do not have the remembered knowledge. These beings, such as the planters, constantly use their minds and their hands to devise and construct artefacts. They are intelligent enough to think out anew the ways of doing things, although they do not remember that these things have been done before. It is as if the whole disease were starting all over again.

Dig a shelter today. Build a house tomorrow. Clear a forest for a city the day after. Choke the landscape with the waste materials the next.

Plant a seed today. Cut down a clearing for many seeds tomorrow. Deforest and irrigate a valley the day after. Change the global climate the next.

Make a spade today. Make a spear tomorrow. Make an explosive machine the day after. Engulf a plain with instantaneous fire and leave it a poisonous ruin the next.

Although the travellers make it their work to frustrate any of this activity wherever they find it, they also use its results. In the far north where they go when times are warm they eat the food grown in the enclosures by the planters. In the far south, when they have travelled there along the spines and ridges of high ground between the foul low-lying slimelands, they eat the roots and tubers stored in the cooled chambers of the hivers. It is a paradox that they do not even try to solve – they are, after all, human beings.

Things are set to change, however. It was not just the making of things and the deliberate changing of the planet that killed their ancestors. The planet itself undergoes changes from time to time, and these changes were such that their ancestors could not withstand them. A force within the Earth that allowed them to tell which was north and which was south died away and then reversed: that was one of the factors.

That same force is used by the travellers themselves; something, some sense inside them, allows them to detect and follow it. Over the past few generations, however, it has been fading away again, and now travel between feeding grounds is going to become increasingly difficult.

The travelling party of 15 contemplate this, as they sit in

The body and limbs of Homo vates, *the seeker, have atrophied from lack of use. Telepathic powers have weakened its other senses and removed its need for eyes and ears. The hivers now feed, protect and carry their guides.*

2 MILLION YEARS HENCE

HIVERS

Alvearanthropus desertus

A harsh and arid habitat has forced the socials to evolve into hivers – all individuality curtailed by the group's need to locate water and food. A hump of fat across the shoulders still provides sustenance in the barren season, while heavy lids now protect their eyes against sand. Longer legs allow the hivers to travel great distances.

the cave mouth, watching the rain hurtling down, stirring up the smells of the forest. This cave, in fact this whole hillside, is unfamiliar to the party. They have never passed it in previous years, so they must have gone well off course. It should not be too much of a problem: once the skies clear they can take their direction from the sun and the stars.

If the skies clear.

Night is falling, and the wet greyness is becoming darker. They are going to have to spend the night here, but at least they have the shelter of the rocky overhang.

When morning comes there are only 12 of them. During the night something has come out of the cave and taken away the other three – something that their communal memory has not anticipated, something with small human-like feet that have left damp prints on the rock.

The survivors move on. The skies are not clear, but they would rather make a guess about which way to go than stay in this place.

HIVERS

The seeker is a tiny, wizened object – a degenerate fragment of its ancestor. It has no need of legs, since it is carried everywhere, and so it has none. It has no need of arms, since everything is done for it, and so its arms and hands are atrophied. It needs neither eyes nor ears, since the only sense it uses is deep down within its head, and has no external organ; so its eyes and ears are sunken and shrivelled. It is merely a head with a nose and mouth, and a little body.

It nestles within the huge hands of the bearer – a sterile adult female that has been turned away from life as a nurse and potential queen deep within the hive and kept at the surface as part of the foraging bands.

The adult males, the warriors, have changed little in outward appearance since the hive communities first evolved. If anything, their legs have become longer, enabling them to cross open spaces more quickly and to forage over large areas. Their bodies have become smaller, and have lost their pot-bellied appearance, since the warriors hardly ever eat grass now and have little need of the voluminous intestinal bacteria vats of their ancestors. The cellulose-cracking enzyme produced by the engineered pancreatic gland is still being produced, but not in such quantities as previously. They eye-coverings are dark, shielded from the harsh glare of the sun, and protected against the stinging sand by heavy lids. The nose is bulbous, the internal passages winding between bony panels covered with a damp membrane that moistens and cools the harsh desert air long

2 MILLION YEARS HENCE

THE HIVE

Alvearanthropus desertus/Homo vates

The hive itself is a massive rock-like structure, with breathing chimneys and thick vented walls similar to those of a giant termites' nest. Flat sloping roofs jut out to provide shade in the heat of the day. Tunnels and shafts beneath the hive reach down deep into the water-table where food is kept cool by constant evaporation from the moist walls. Damp air from the lower levels is driven through the hive by wind movement across external vents.

The queen is protected and provided for in caverns deep below the hive. Food is gathered for her by the young hivers. Warriors guard the ancient hive and her person. Nurses feed her young.

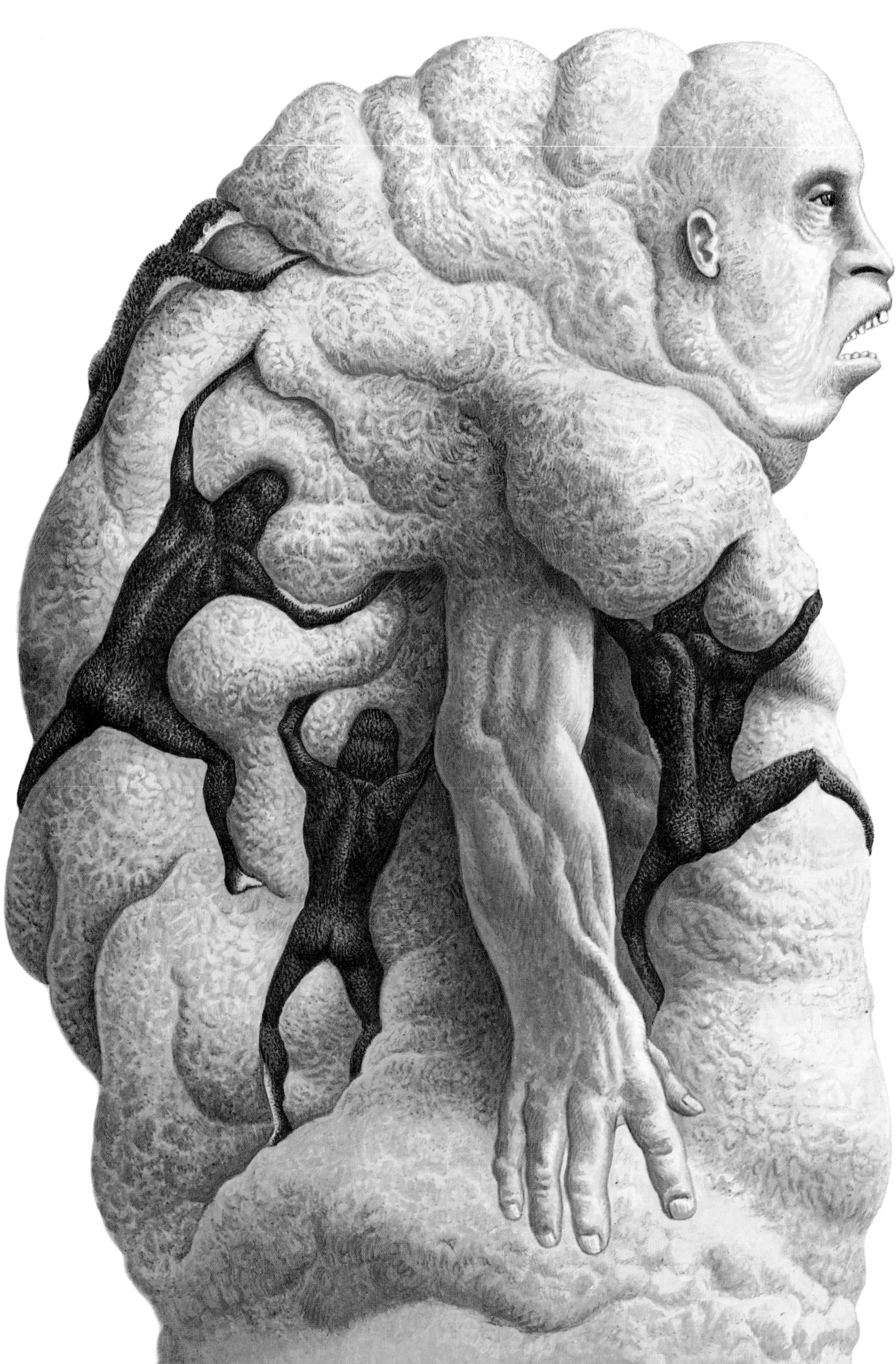

2 MILLION YEARS HENCE

HOST/PARASITE

Penarius pinguis/Nananthropus parasitus

The islanders have evolved parasitic feeding habits that rely on the tundra-dweller's metabolic need to produce surplus fat. In this way, the obese tundra-dwellers have found an ecological niche that allows them to exist now that the tundra plains have disappeared and the mountain tribes failed.

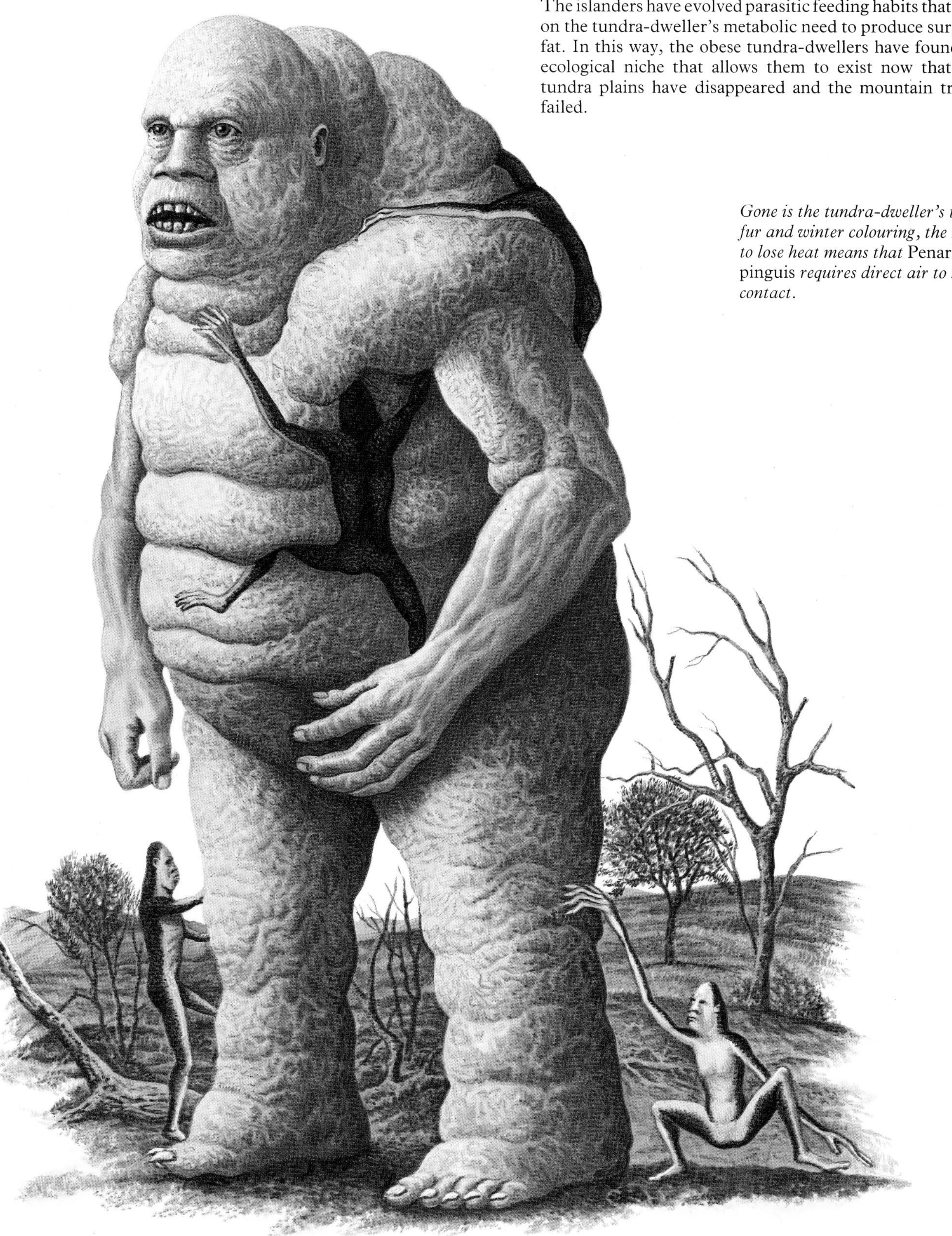

Gone is the tundra-dweller's thick fur and winter colouring, the need to lose heat means that Penarius pinguis *requires direct air to skin contact.*

before it reaches the lungs. A bushy moustache around the nostrils and across the upper lip filters the grit and dust from the breathed air. A smooth hump of fat over the shoulders and neck is established in the wet and abundant season, but this tends to shrivel away when the climates become dry.

It is mostly in their behaviour that they differ from their ancestors. Now they have no individuality at all, listening for the few grunts of command from their leader and obeying blindly. It is not in the interest of the hive as a whole for anyone to show an individuality, and so it was lost generations upon generations ago. Now and again, however, it surfaces once more, and under the influence of these throwbacks hives begin to experiment with new and different ways of living, which nearly always end in failure. The progressive hive dies, turns to dust, and the neighbouring hives absorb its territory.

As always, the youngsters, male and female, make up the gathering parties, using their big hands to dig in the soil and carry the food that they find. When they come of age, the males develop into warriors, and eventually may become breeders. The females become nurses, with the possibility of becoming queens some day; or else they become bearers, entrusted with the task of satisfying every need of the all-important seekers.

This day is much like any other. The party of gatherers, guided by the seeker and guarded by the warriors, sets out from the hive in the pre-dawn, the coolest time of the day and the best for travel. Behind them, a silhouette against the lightening sky, lies the bulk of the hive; its flat roofs jut out like natural rock formations to produce the shade in the heat of the day, the vertical walls beneath the overhangs form banks of variously-sized openings for access and ventilation, and its many chimneys and breathing funnels point up like fingers and arches against the sky.

Deep below is the maze of passages and chambers dedicated to the housing and comfort of the queen and her young offspring. Here lie the food storage units cooled by the constant circulation and evaporation of water from moist walls. The dampened air is then carried around the hive through the living quarters by an ingenious network of finely-fashioned holes and tunnels, driven by the natural movement of the wind across the external vents. The vapour is eventually recondensed to liquid before the stale air is lost to the outer atmosphere. The water for all this is brought up from the deep wells and waterpits by capillary action through the rocks.

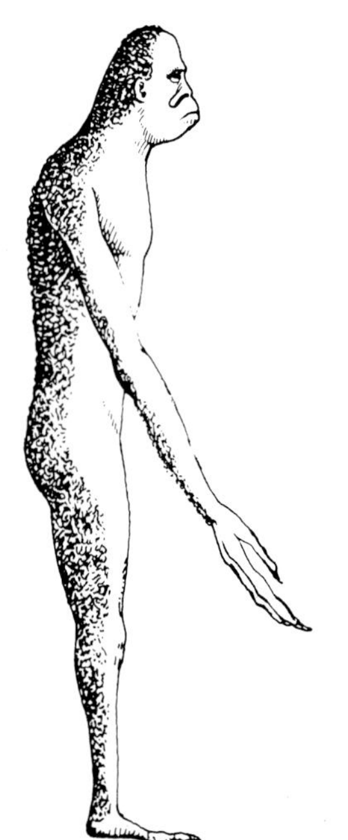

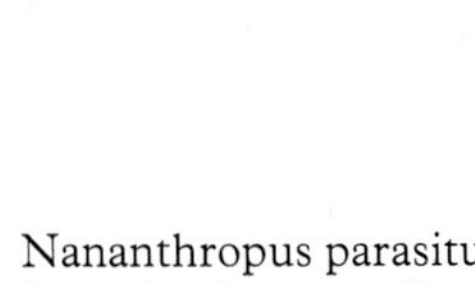

Nananthropus parasitus *have developed small blood-letting front teeth.*

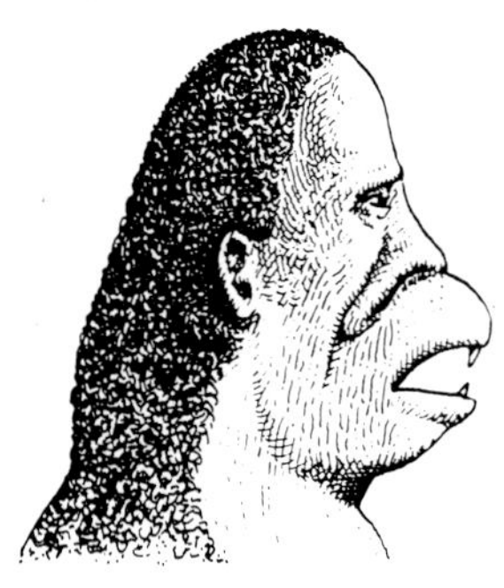

The only function of the long fingers and toes is to allow the parasites to grip folds of fat.

tundra-dweller
aquatics
seeker
hiver
islander
parasite
host
fish-eater
antmen
spiketooth
desert-runner
slothmen
tree-dweller

Homo sapiens sapiens
Homo aquaticus
Homo caelestis
Homo sapiens machinadiumentum
Homo virgultis fabricatus
Homo glacis fabricatus
Homo silvis fabricatus
Homo campis fabricatus
Piscanthropus submarinus
Homo sapiens accessiomembrum
Homo mensproavodorum
Speluncanthropus
Moderator bauli
Baiulus moderatorum
Homo dormitor
Homo vates
Alvearanthropus desertus
Homo nanus
Nananthropus parasitus
Penarius pinguis
Piscator longidigitus
Formifossor angustus
Acudens ferox
Harenanthropus longipis
Giganthropus arbrofagus
Arbranthropus lentus
Piscanthropus profundus

The party, 100 strong, takes its usual route along the undulating foothills, skirting the dreadful slimelands on the right, and the barren rocky uplands on the left. Beyond, the slope widens out into a valley in which water flows for much of the year, and where plants can grow and there are usually tubers or thick roots to be had.

Before their narrow path widens the leader of the party grunts an order to halt. The seeker is agitated, but is not telling them that there is food close by: it is telling them that others approach.

With another grunt the leader calls the warriors together in a protective wall; but they need not have worried. Those who approach pose no threat.

It is full day now, and the party can see five or six shambling creatures moving down the rocky slope towards the slimelands. The bodies are bulky (very bulky for the size of their legs) with thick hummocks and rolls of fat seeming to engulf them. Dull faces look out from the folds of pale flesh. In the dim light, however, the parasites are just visible: tiny and spider-like, four or five of them are embedded in the deep fat of each figure, their faces buried and unseen, feeding continually from the creature's surplus.

No threat to the hive, and so of no interest to the party; but the leader does recollect that more and more of them are seen nowadays wandering over their domain. They seem to be spreading from the forest areas that are their home. Dimly the leader wonders what they find to eat here, and how they protect themselves from the harsh sun. He does not wonder for long, however. With a backhanded gesture, he brushes the first of the day's sand out of his moustache and signals for the party to move onwards. Soon he has the party on the move once more and the strangers have been completely forgotten.

Had the party stayed to watch, they would have observed the lumbering creatures scramble down into the flats of the slimelands and wade out amongst the disgusting blue-green sogginess. Dumbly they scoop up handfuls of the slime, exposing the yellow stench beneath, and begin to feed on it. The parasites embedded in their fat ignore all this. The food, be it nuts, leaves or slime, will be converted into huge deposits of fat and tissue that will sustain them.

The parasites and their hosts are not the first communal creature to arise since the days of the engineers, but they are the only surviving type. The symbionts, in which the hunters teamed up with the tundra-dwellers, to live on the cold plains, are extinct now. They took to the mountains after the cold plains faded away, and there they existed for some time; but they were never really developed as mountain creatures, and all kinds of maladaptations began to show themselves. Eventually the populations dwindled and the whole race died out.

That is not the case amongst the parasites and their hosts. The hosts, too, are descended from the tundra-dwellers, but unlike the carriers of the symbionts they changed as the conditions changed. Gone are the woolly coats and the resistance to extreme cold, but they still retain the thick deposits of fat. Indeed their metabolism generates more fat than they could possibly need, and that is what sustains the parasites. The energy and raw materials for all this production comes from the constant consumption of plants – any kind of plants, including the blue-green algal cultures that the aquatics developed as their own food source and spread over the lowland areas of the globe, turning them into the foul slimelands so despised by most of the land-living creatures.

It is not only the hivers that ignore the parasites and their hosts as they wade into the featureless slippery mat. Also ignoring them are the aquatics, not far away, looping and slithering about in the moist yellow depths below the slime crust. They are grazing their way through the algal culture that their ancestors established aeons ago on the lowlands above the surface of the ocean. There is plenty of food for them now, not like in the days of want. They know very well that some creatures from the land come and steal from the edges, but the losses are small. The only trouble is dehydration. If the algal covering is breached there may be a considerable water loss before it has a chance to grow again; but with all the world's lowlands covered in the self-sustaining food-generator there is little to worry about.

3 MILLION YEARS HENCE

FISH-EATERS

The brook burbles down the slope, bouncing off the exposed rocks and rubble in the gully, washing soil from the banks beneath the hanging tangled roots of the great deciduous trees. Newly-hatched flies weave and gyrate in the cool sunlight above the little pools and backwaters that gather beneath and behind the waterfalls. The exposed rocks are pocked by smooth circular potholes, worn by the

The eyes of Piscator longidigitus *polarize light, removing the bright reflections that normally prevent animals seeing below the surface of water. His brain automatically compensates for the refraction.*

3 MILLION YEARS HENCE

FISH-EATER

Piscator longidigitus

Three million years have passed and the results of constant natural selection and evolution are apparent. The temperate woodland-dwellers have diversified, and developed specialized body forms to fit different environments.

Living by upland lakes and beside rivers, the fish-eater is equally at home on land and in the water. His pelt is smooth and glossy, his shape streamlined. Ears are small and close to the head, the neck is short and feet are broader than usual.

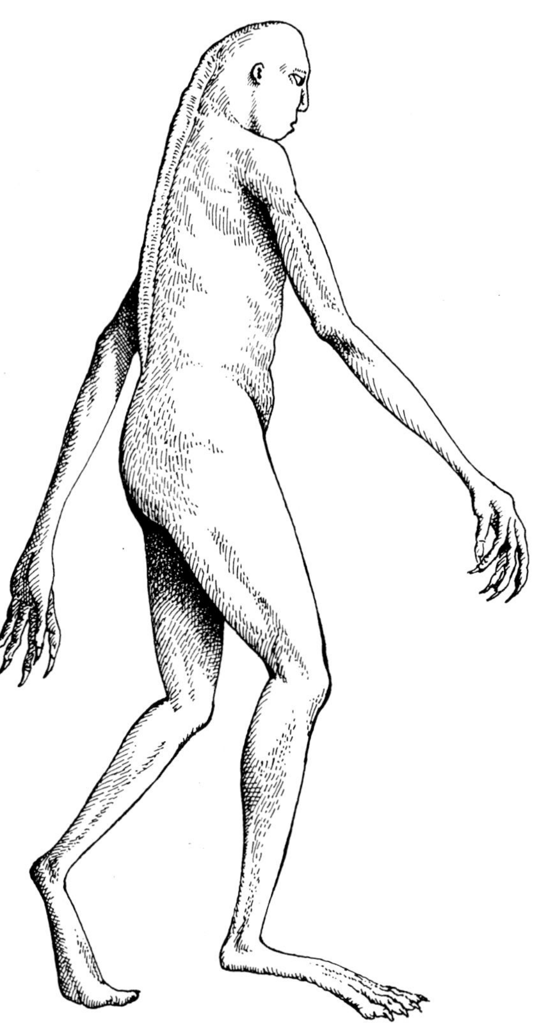

The fish-eaters have evolved by natural selection the streamlined shape earlier engineered into the aquatics.

swirling stones caught up in the infrequent floods. At present, though, the stream is flowing with gentle splashes and gurgles, through the V-shaped cleft in the soil, and downwards through the wooded hillside towards the distant plains.

The air is cool, almost as cool as it was during the ice ages of ancient times. There can be no more now for a very long while. The continent at the south pole is covered with ice, but there is no permanent icecap in the north. The gradual movements of the continents has opened the oceans to such an extent that warm currents from the equator now sweep up to the polar sea and keep it permanently free of ice.

There is less carbon dioxide in the atmosphere than there has been for a long time, and this is the reason for the cooling. Sunlight shining down onto the Earth's surface is re-radiated away into space, with little of it being trapped in the layers of air. The algae that were induced to grow on the lowlands by the aquatics have absorbed much of the atmospheric carbon dioxide, and now it lies trapped in vast deposits of peat and lignite below the roots of the forests of the coastal plains. The aquatics themselves have long ago abandoned that wasteful exercise, and now grow more concentrated food out at sea.

In the shadow beneath an overhang, screened by the interwoven arches formed from the sturdy roots of a great tree and by the more delicate soil-clogged roots of the grasses and undergrowth plants, there sits a figure. If he had the wit to interpret them, the rocks in the bank behind him would tell him of an important part of his history. They are normal strata of dark finely-bedded shale, except for one thin layer which is quite unusual. Shale is formed from compacted mud that was once deposited layer upon layer in quiet waters, but this one particular bed of the sequence seems to consist of a different material altogether. It looks as if all kinds of foreign matter spread in and were deposited on top of the mud at one particular time. It is a very thin bed, and so the deposition could only have taken place in a period of a few thousand years at the most. The top boundary of this layer is as abrupt as the bottom, and above this the normal sequence of shale continues, showing the continuing deposition of clean mud. Evidently the continuous deposition of mud in the area had been disrupted for a short period while great changes took place in the world at that time, and the resulting bed of foreign matter had eventually been turned to rock along with the mud above and below.

The figure has never noticed this. It is not part of his life

The hand has evolved two strong fingers that allow the tree-dweller to hang from the underside of branches.

3 MILLION YEARS HENCE

TREE-DWELLER

Arbranthropus lentus

The long hooked fingers evolved to cling to the jungle canopy, but they can also break open insect nests under the bark. Small but slow, the tree-dweller moves with deliberation through the humid rainforest, clinging tightly to the underside of the great spread of branches. Fruit is plentify and insects abound. With no enemies and abundant food there is no need for speed, aggression or change. Without the need to adapt or develop, the sloth-like tree-dweller will remain in a state of statis, able to breed but unchallenged.

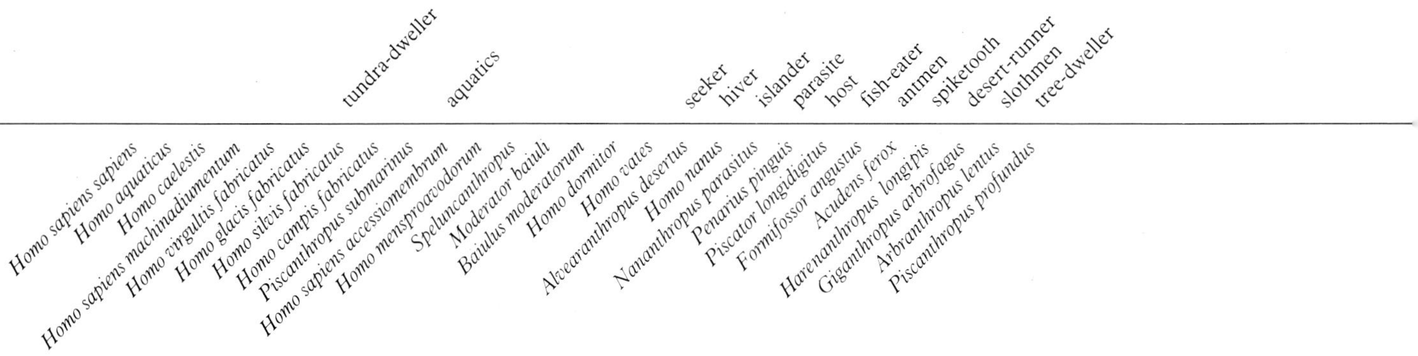

tundra-dweller aquatics seeker hiver islander parasite host fish-eater antmen spiketooth desert-runner slothmen tree-dweller

Homo sapiens sapiens, *Homo aquaticus*, *Homo caelestis*, *Homo sapiens machinadiumentum*, *Homo virgultis fabricatus*, *Homo glacis fabricatus*, *Homo silvis fabricatus*, *Homo campis fabricatus*, *Piscanthropus submarinus*, *Homo sapiens accessiomembrum*, *Homo mensproavodorum*, *Speluncanthropus*, *Moderator baiuli*, *Baiulus moderatorum*, *Homo dormitor*, *Homo vates*, *Alvearanthropus desertus*, *Homo nanus*, *Nananthropus parasitus*, *Penarius pinguis*, *Piscator longidigitus*, *Formifossor angustus*, *Acudens ferox*, *Harenanthropus longipis*, *Giganthropus arbrofagus*, *Arbranthropus lentus*, *Piscanthropus profundus*

and he is looking the other way. Sunlight, sparkling from the pool below him, casts ripples of light on his face and arms. He has the long limbs and the long face of one of the hunting people, but there is something a little different about him. His neck is shorter, his ears smaller, his fingers longer, and his feet broader than usual. Also, his eyes are strange – not in their appearance but in their function. The lenses smooth out the bright reflections from the water's surface enabling him to see directly into the depths. His brain compensates for the refraction and distortion caused by the different densities of the water and the air. He uses these faculties to watch the bottom of the pool for his prey, for this creature feeds on fish.

In the temperate regions of the world, where the forests and woodlands still exist on the upland slopes, the hunting people still pursue their age-old lifestyle, just as they have done since they were engineered. However, as there are so many different food sources in the habitat, many of them have begun to specialize, and to develop bodily forms that are appropriate to their particular way of life. Most lie in wait for birds, or dig in the ground after burrowing mammals. Some even feed on nothing but insects that they remove from the layers of their wooden homes.

One group has developed as an almost exclusive fish-eater. Living mostly by the hilly lakes and rivers, these creatures spend most of their time on dry land, but enter the water to chase their prey. Their broad feet help them to swim, and their long fingers can spear their slippery prey with ease. Their pelt has become particularly smooth and glossy, and they are beginning to adopt a streamlined shape to their bodies, with a bulbous head tapering into the smooth shoulders without much of a neck. Their eyes work best above the water, but their focus can be adjusted to allow their use beneath the surface as well.

The individual beneath the overhang – so still that he appears to be asleep – suddenly focuses his eyes on a movement not far below the surface of the pool. A long fish swims in from the more turbulent area near the current, its deep tail whisking back and forth, moving its body lazily along with an ease that would make the watcher feel jealous if he had the capacity to feel such emotions. Taking his time, he watches the creature come closer and closer.

His hand cleaves the water so expertly that it hardly makes a splash. The pointed claws on the long fingers close around the scaly body, and pin it before the slippery shape can wriggle free. Then, with an almost reflex jerk, he yanks it from the water and onto the bank beneath the overhang. With a swift blow he kills it.

Then he eases himself from his hiding place, straightening out the slight cramp in his muscles, and gathers up his catch to take it back to his mate and family.

No, he is not a fully-adapted water creature. There are other derived humans in the world who are more perfectly built for the water environment. Nevertheless he is good enough to survive and to continue his line.

TREE-DWELLERS

Far away, on another continent, a much smaller creature moves slowly, upside down, through the dripping branches of the tall trees. Her fingers and toes are permanently curved, and allow her to hang on the underside of the stoutest branches.

Slowly she turns her little head and looks about, seeking out the next piece of food growing in the humid air. There, on the next tree, is a bunch of fruit. Carefully she crawls along beneath the branch back to the trunk, where she can climb out amongst the branches closer to it. Dimly she sees that there is another creature, a male of her own kind, already on that tree, well above the branch with the fruit. He is moving slowly downwards. Whoever reaches the fruit first will claim it.

Her long fingers reach for the next hand-hold, and splinter through a weak thin layer of bark. The air is suddenly full of noise and aggression. A cloud of insects has burst from the hole and is thudding into her, jabbing through her pelt with pointed tail weapons. She feels the prick of the attacks, but there is no pain, as her line became immune to the poisons generations ago. She knows that there is good eating here, so ignoring the insects that are swarming and clustering around her hands she breaks up the bark covering the nest that she has disturbed. Combs of honey and grubs are stacked in there, vertically in neat rows. With her usual deliberate actions she breaks them from their hollow and chews contentedly.

Afterwards, with the nest empty, and the insects spent, exhausted or fled, she remembers the fruit on the next tree. With painfully slow movements she unwinds from her feeding position and begins to crawl along the branch once more.

Eventually she comes in sight of it, but she is too late. The male has already reached it and is eating. No worry. She has fed, and there is plenty of other sustenance around. She turns to crawl away again; but then she stops because the male has noticed her and is crawling along the underside of the thin branches towards her.

He obviously wants to mate. Does she want to let him? Yes, this is a good time since they have both eaten and will have the energy. It is also a long time since she gave birth, and her child has now matured and left, so she can take on the responsibility once more. Meekly she awaits the male's approach.

The rainforests that still clothe the windward mountains of the moister parts of the globe and the great river basins along the equator still have tree-dwellers, which in most places have changed little over the millennia. The long arms and long-fingered hands that grasp branches allow them to hang firmly onto their high perches. The long legs with the prehensile toes allow trunks and boughs to be negotiated. The weak intellect that knows only about food and mating, and about those only enough to satisfy the basic drive for existence, allows the creature to survive. Food has always existed here, and, seemingly, will do so for ever; therefore the tree-dwellers have no need to change, unlike the creatures indigenous to the other habitats of the world.

The only change has occurred in the pace of their lives. With no enemies, the tree-dwellers in many areas have become slow and ponderous, moving sluggishly from one meal to the next, from one mate to another. There is no strife, either with one another or with different types of creature. Perhaps someday, when something unforeseen comes and takes away the forest, then perhaps the tree-dwellers will alter. That is, if they still have the genetic capacity for adaptation, if they have not lost the inherent ability through a long period of stasis and inbreeding.

Any change to the environment, however, will not take place for a long time yet.

ANTMEN

A long finger probes and gropes down the tiny tunnel into the nest. The loose soil and twigs are forced apart by the blade-like fingernail and the finger slides in, deeper and deeper. Ants, enraged by the intrusion, swarm out of side-chambers and tunnels, and mass against the attacker. Stings and jaws sink into the tough skin, but make little impression. Courageous fighters hang onto the invading flesh as their blind instincts dictate, while others climb over them to find other spots to attack. Soon the whole finger is a clump of swarming defenders.

Up above, the antman has gauged that enough time has elapsed, and pulls his hand with its long finger from the nest. It is a black mass of ants. He has judged the timing correctly – just enough time for the ants to attack his finger in sufficient numbers, but not enough for them to abandon the defence as useless. He did not feel the assault on his finger, since it has no nerves that would detect pain. The whole finger, with its attached ants, goes into his mouth and is then withdrawn slowly, his tiny teeth scraping the insects from the skin. He swallows the ants, a number of which saw the danger in time and abandoned the finger, and are now crawling over his face. They do not trouble him: he can close off his nostrils and his eyes as they come close, and when his mouth is empty he wipes them from his face with the back of his hand and his long tongue.

He turns back slowly to the nest. With the huge claws on two of his fingers (those that were once called the thumb and index finger) he rips the covering off another part of the nest. Patiently he waits for the defenders to swarm up once more, and inserts his long middle finger again into one of the passages.

He is rather a solitary creature. The ants that he eats are highly nutritious, but it takes a great deal of them to make a meal so a single anthill could hardly sustain two antmen. His movements are also very slow and deliberate. He has no natural enemies, although he evolved at the same time as many of his cousins developed into hunting, flesh-eating types. His defence is in the food that he eats. He is immune to the poisons of the formic acid in the ants' stings, but his body does not break them down; instead it redeposits them in his tissues, making his flesh unpalatable to any meat-eater. His fine black fur has a glaring white stripe across the back and down the legs. Any meat-eater that sees this striking pattern realizes that its owner is not good to eat.

Once upon a time, millions of years ago, there were other animals that pursued this very way of life. They inhabited all the continents, but each place had its own unrelated species. The anteaters of old South America were no kin to the aardvarks of Africa, and they only looked like one another because they pursued the same lifestyle. They possessed similar bodily features that had the same functions – long sticky tongues, narrow mouths, heavy claws – but evolved independently. Likewise neither of these animals was related to the marsupial numbat of Australia, an ant-eating animal of similar appearance. The whole concept of the same shapes cropping up in unrelated animals that lived in the same way was what the zoologists once termed 'convergent evolution'.

Now all the anteaters, the aardvarks and the numbats have been extinct for 3 million years, yet their food has remained: there are still ants and termites all over the world. It is the way of nature that if a food supply exists then a creature will evolve to exploit it, usually emerging from a group of fairly unspecialized animals. In this case, the most unspecialized animals around were the humans genetically engineered to live on the wide range of food of the temperate woodlands. Consequently, over the last few million years these omnivores have developed, under the natural influences of selection, to become specialized feeders in the various different environments present. One group has developed into the anteaters.

DESERT-RUNNERS

The sun burns blisteringly down, baking all the landscape and beating up from the sharp naked rocks and the pockets of dry dust that lie between. All is yellow and grey, and no plants are to be seen anywhere. In a wadi (a gushing torrent in the distant rainy season but now a parched gully) the sand lies deep and barren. The only sound is the distant hum of the wind, and the constant hiss of sand as it is blasted against the rocks and swirled about in the hollows. The monotony is broken by a faint scrabbling sound as a brown lizard scuttles amongst the loose stones and vanishes into their shadows, then all is stillness again. Few things venture out in the killing heat and dryness of the desert noon.

Yet in the distance something large is moving, and moving quite swiftly too. Its legs and arms are long and thin, and its head seems inordinately large, covered in white hair and surmounted by a pair of huge ears. It looks like one of the hivers, but it is travelling and hunting alone. It is, in fact, one of the hunters that has evolved and adapted to the harsh conditions of the desert – a desert-runner.

His long strides take him swiftly across the scorching wadi and into the sharp blackness of the rock shadow at the other side. There he rests, looking out at the dazzling sand with his polarized dark-lensed eyes. He sees things only in black and white, as the rod cells of his eye have developed at the expense of the cones, increasing his distant and night vision. He has just travelled many miles over the rocks and dust and will now rest a while to cool his body. Despite his adaptations to life in the desert he must still guard against

3 MILLION YEARS HENCE

ANTMEN

Formifossor angustus

Some diets are so specialized that the entire body form evolves to accommodate them. The slow-moving and solitary antman has claws for tearing open anthills, a long middle finger for reaching into the tunnels, and a startling coloration to warn enemies that its flesh is not good to eat. Extreme adaptation has lost *Formifossor angustus* the sharp teeth and nails of his woodland-dwelling ancestors. Instead the antman's defence is his vivid coloration and his specialized diet.

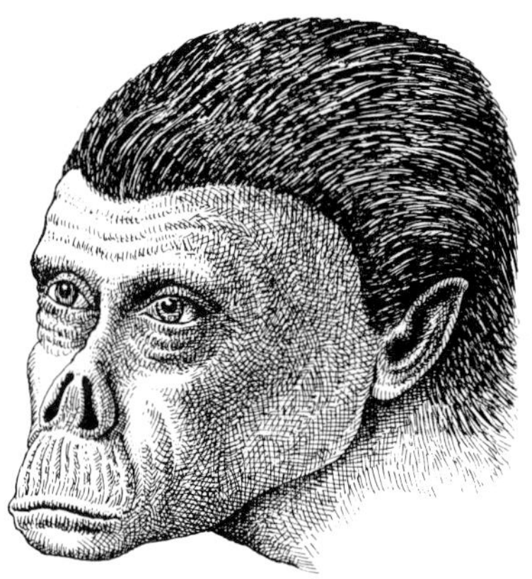

Eyes and nostrils can be closed off against ants. The tiny mouth scrapes swarming ants from the long middle finger.

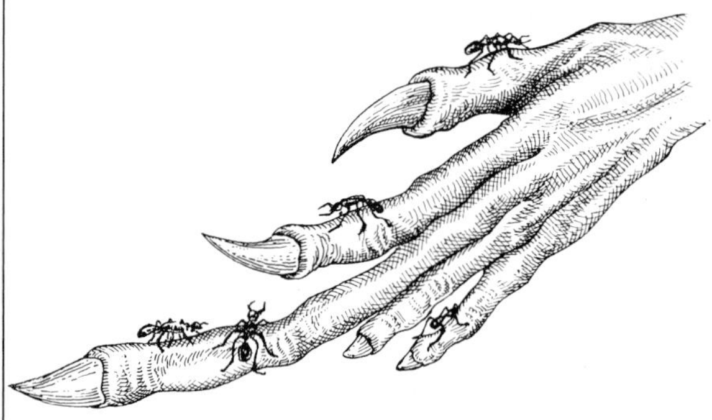

The blade-like nails can cut open anthills. The bony fingers lack nerves that carry pain.

The antman is immune to formic acid, the poison carried in an ant's sting. But his body does not break down the poison, it redeposits the acid in his tissue, making the antman unpalatable to his potential enemies.

3 MILLION YEARS HENCE

DESERT-RUNNER

Harenanthropus longipis

The runner's long strides take him swiftly across the scorching wadi and into the sharp blackness of a rock shadow. There he rests, looking out at the dazzling sand with his polarized dark-lensed eyes.

The fatty deposit across his shoulders is depleted but not yet exhausted. Bat-like ears radiate waste heat. Although similar in appearance to a hiver, the desert-runner's ancestors originated in the temperate woodland. Through convergent evolution, the desert-runners are beginning to adopt the shape that was designed into the plains-dwellers all those millions of years before. However, the runners are carnivorous, unlike the hivers.

the killing heat of the sun and the dryness of the wind. The fatty hump behind his neck is almost depleted, the store of fat having been turned into energy. Nevertheless, he knows that he will soon reach a fertile spot where his stores can be replenished.

The hump is only one adaptation to the dry heat of the desert. Greatly-enhanced kidneys distill and use every drop of water that enters the body. Waste heat is radiated from the body by means of the great bat-like ears, acting as heat-exchangers, and the long thin legs that give a very large surface area for the mass of the body. These are necessary, since no sweat exudes from the skin – all water is saved. The ears, eyes and nostrils have thick folds of skin that can close them off and keep out sand and dust when the winds get too high.

The sun has now passed its zenith, and the black desert shadows are lengthening. Rested now, the desert-runner creeps out to continue his journey. The first part of his travel was over sand, where he used his long legs, with their light elongated foot bones powered by the concentration of muscles in the thigh. Now his way takes him over naked rock, so his passage is slower, using his long toes and gripping fingers to find purchase in the cracks and joints of the hot crumbly stone. As the sun descends into the dusty haze of evening, his goal is in sight.

The hive looks like one of the rocky hills that surround it. Its vast roof slabs look just like the horizontal strata of the surrounding rocks, and the black entrances just like the wind-blasted caves of the dusty crags. Just as desert-living humans evolved along the lines of the desert animals, the desert cities of the hivers developed along the lines of their habitats. The vast thick roofs paralleled the flat stones that absorbed the heat of the sun and protected the creatures that existed underneath. The tunnels burrowed deep into the Earth, cool by day and insulated from the bitter cold of the night. Water was gathered by vast dew-traps in the surrounding sands, and food was gathered from wide areas and brought swiftly to the cities by the foraging teams.

The desert-runner will spend some time here. The hivers eat only plants, while he eats only animals, so they will not conflict with each other. The moisture that is generated in and around a hive and the food stored within attract all kinds of insects, reptiles and small mammals which the desert-runner will hunt, while the hivers, with completely different nutritional requirements, will tolerate his presence.

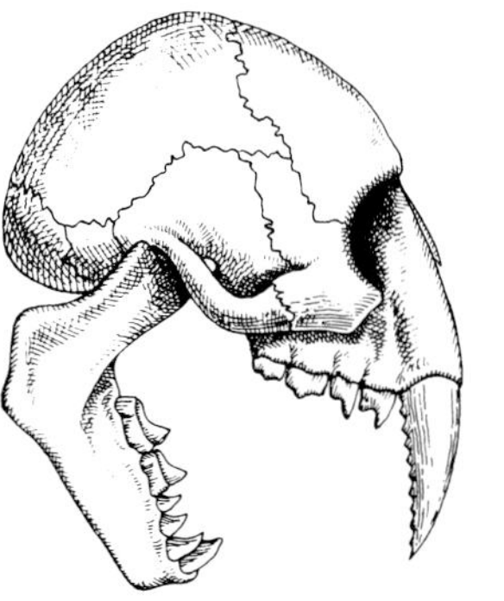
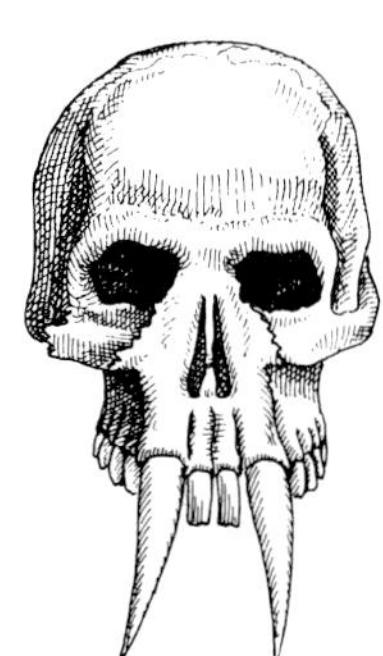

The slashing teeth of Acudens ferox *have evolved from the incisors of his original ancestor,* Homo sapiens.

SLOTHMEN AND SPIKETOOTHS

A huge mound of fur-covered flesh, an indistinct lump in the luxuriant undergrowth, pushes its way through the bracken and brambles. It makes contented burbling noises to itself as it goes. Insects, small mammals and birds burst from cover to get out of the path of the great creature as it crunches slowly through the greenery. It is quite harmless, but its immense weight causes a great deal of damage as it passes.

It stops by a tree and looks slowly upwards. There are appetizing fresh green leaves up there. Using its forelimbs against the trunk, it slowly pulls itself upright. Now it begins to look more like a human being, to be precise the tundra-dwelling human being, that was its distant ancestor.

From the bushes beyond, a number of other creatures stop feeding and move out of the way. They are also descendants of the tundra dwellers and have grown large, but not nearly as big as this great creature. Nor have they changed much in the last million years or so: they still produce the huge quantities of superfluous fat, and are still infested by the tiny parasites that live on the excess.

The tundra-dwellers that adapted to woodland life did so very successfully. Their heavy bodies were well supplied by the voluminous plant life of the habitat. Evolution had produced the right shapes by trial and error; man copied them, and then evolution took the copies and modified them further. If these big creatures have a parallel

In carnivores it is normally the pointed canines that develop as killing teeth. The spiketooth, however, has a jaw that drops down to allow the teeth to be used efficiently, and it is the upper incisors that have become the weapon.

THE SPIKETOOTH

Acudens ferox

Large plant-eating animals inspire the evolution of meat-eating creatures to feed on them. *Acudens ferox* is heavier than other hunting species. It can afford to be, needing neither speed nor stealth to hunt the slothmen. It has slashing front teeth able to penetrate the thick fur and tough skin of its prey.

Although much larger than the tundra-dweller, the slothman retains the proportions of the species from which it evolved. The fat layers are still in place and heavy claws are needed to pull the huge body upright.

3 MILLION YEARS HENCE

SLOTHMEN

Giganthropus arbrofagus

Temperate climates encourage the evolution of large creatures, bulk retains body heat and large leaf-eaters can find enough nourishment to support their mass. By a process of convergent evolution the slothman is now similar to the giant ground sloth of South America from pre-human times. But two factors were needed to allow the tundra-dwellers to evolve into slothmen – plentiful food and no enemies. Sustenance is still there but now they face a newly-evolving predator.

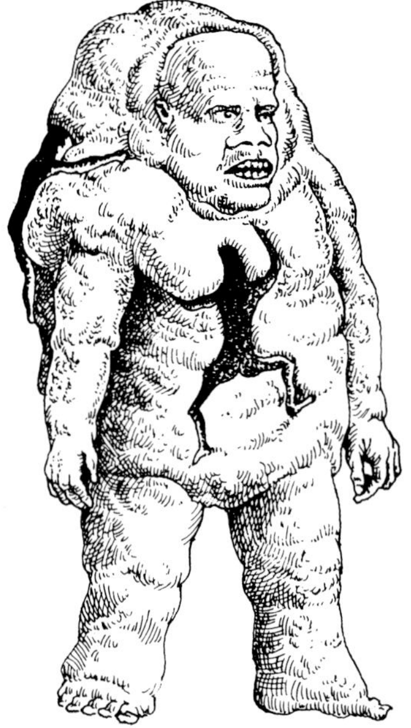

Tree sloth form, parasite host with parasite and spiketooth. All come from the same basic stock.

– a convergence – with any creature from the fossil past it would be with the giant ground sloths of ancient South America. Like these, firstly, they developed successfully, even with their great bulk and sluggish habits, because there was the food supply to sustain them and they had no natural enemies; secondly, they spend most of their time on all fours, so that their bulk can be well supported, but they can also rise to their hind legs to feed from tall trees; and, thirdly, they have become about three times as tall, and so about ten times as heavy, as their ancient ancestors.

Like the giant ground sloths, too, they are succumbing to a newly-evolving predator.

The hunters have been evolving into many specialized types, each one hunting a specific type of prey: some hunt birds, some hunt small mammals, some hunt fish. One, however, has evolved to hunt the descendants of the big tundra-dwellers. The spiketooth is larger and heavier than the other hunters, not needing stealth or speed for its hunting since its prey is large and slow-moving. What it does need, however, is a specialized killing weapon, and this it possesses in the shape of its front teeth.

Amongst the traditional carnivorous mammals, of which there are only a few small species left, the killing teeth were normally the pointed canines. In extreme types, like the sabre-toothed cats, they developed into long slashing blades that were able to penetrate the thick hides of very large animals. In the spiketooth the weapons have developed instead in the incisor teeth at the front, rather like the only remaining teeth of the parasites that also feed on the flesh of the descendants of the tundra-dwellers. The spiketooth's mouth is very large, allowing its jaw to drop clear of the upper teeth so that they can be wielded efficiently. The hands are large and powerful, with strong fingernails that allow the spiketooth to hang onto the fur of the slothman while it stabs at the neck, or onto the fatty rolls of the parasitehost while it slashes its way through the blubber.

This may seem like cannibalism, since both predator and prey are descended from human beings; but their common ancestor existed so far back in time that the creatures involved now comprise entirely different species. The preying of one upon the other is merely the natural result of the development of a stable ecological system.

The slothman munches placidly at the leaves and twigs, unaware of the approaching danger. Away below him in the undergrowth the parasitehosts have already left, their dim wits sensing the approach of a pair of spiketooths. If the distant crashing caused by their lumbering flight

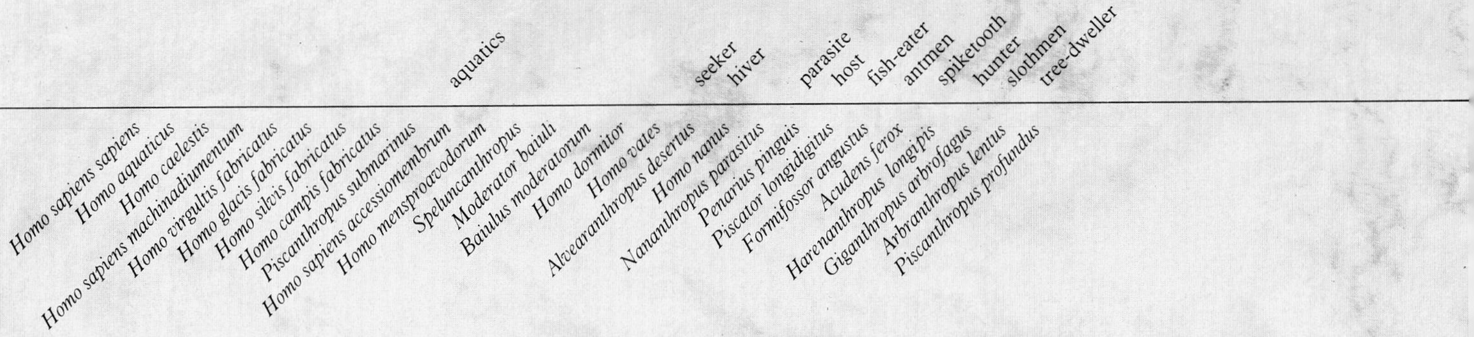

aquatics
seeker
hiver
parasite
host
fish-eater
antmen
spiketooth
hunter
slothmen
tree-dweller

Homo sapiens sapiens
Homo aquaticus
Homo caelestis
Homo sapiens machinadiumentum
Homo virgultis fabricatus
Homo glacis fabricatus
Homo silvis fabricatus
Homo campis fabricatus
Piscanthropus submarinus
Homo sapiens accessiomembrum
Homo mensproavodorum
Speluncanthropus
Moderator baiuli
Baiulus moderatorum
Homo dormitor
Homo vates
Alvearanthropus desertus
Homo nanus
Nananthropus parasitus
Penarius pinguis
Piscator longidigitus
Formifossor angustus
Acudens ferox
Harenanthropus longipis
Giganthropus arbrofagus
Arbranthropus lentus
Piscanthropus profundus

through the thickets causes any concern to the slothman, he does not react to it. He does not react at all until the familiar form of a spiketooth steps out from the shade of the forest and he suddenly recognizes the shape and the smell. Slowly he turns away from the tree, turning his back on his enemy, and begins to descend to all fours.

The first spiketooth, less experienced than the other, leaps for the broad back, hooks onto the long fur, throws up his head and drops his jaw ready for the strike. This is a mistake, as it enables the slothman to use his only weapon – his weight. He slowly topples backwards, while the attacking spiketooth tries frantically to untangle his claws from the fur. Remorselessly the attacker is pressed back down into the bracken and the soil of the forest floor, and the slothman lands spreadeagled on his back with his enemy crushed to death beneath him. However, this makes him vulnerable to the spiketooth's mate. She now leaps upon the unprotected chest and plunges her long killing incisors into the slothman's neck.

The kill is a success, which is all she knows. There is no grief for her dead mate. The spiketooth has evolved so far from the original human state that she feels no emotion at all.

5 MILLION YEARS HENCE

MOVING STARS

Strange stars move in the frosty sparkle of the night sky. The eternal star patterns themselves have moved little in 5 million years, but now there are new stars superimposed upon them; and these stars are in gentle continual motion. They are ignored by the creatures below, who do not appreciate what profound changes are about to be inflicted upon their world.

The sluggish beings in the trees of the forests can comprehend nothing higher than the tops of the trees that they inhabit. Nor do they need to, all their food existing to hand in their own environment. Nothing outside can have any effect on them.

Out in the vast deserts, the members of the few huge hives that are left continue their lives in the familiar mechanical way. Permanently-manned foraging and gathering routes reach out, tentacle-like, from the massive hubs that consist of labyrinthine subterranean bunkers, all swarming with ordered and predestined life. Amongst these millions of individuals there is not one mind that can comprehend the heavens, let alone the significance of new moving stars.

The decadent parasites, embedded in the fat layers of their grotesquely misshapen hosts, care nothing beyond their hosts' continuing survival; and their hosts are mere feeding machines, dumbly eating, eating, eating.

The swift hunters, specialized for catching birds, small mammals, fish – or even the parasites' hosts – may wonder about the movements in the night skies above them; but they have not the wit to imagine that these events could possibly have any effect on them.

Out in the oceans, the teeming aquatics know little of what happens above their watery ceiling. They can hardly comprehend the existence of life on land, let alone the nature of the stars in the sky.

Only the possessors of the hereditary memory could have understood, but these have been extinct for millennia. Their religious refusal to use the knowledge that they all possessed meant that they could do nothing to help themselves to improve their situation. When natural conditions changed they refused to change as well. The Earth's magnetic field reversed, continents moved, and changing sea levels cut off migration routes. Rivers changed their courses, volcanoes threw up new barriers, and climates altered from year to year. Creatures of lesser wit and no knowledge of the past survived these upheavals, which constituted disasters on a local scale, but merely inconveniences on a global one. However, amongst those with the memory, the changing conditions took their local environment further and further away from what they knew or remembered, and eventually, rather than change with it, they perished.

The coniferous forest is black and silent in the night. Hunters lie huddled, asleep. The trees jut up black spikes into the sparkling sky – the sky in which there are now, for the first time in 5 million years, slowly-moving particles of light. Overhead a star, one of the new moving ones, is glowing brighter than the rest. It expands and descends in a gentle arc across the sky, stringing behind it a fading trail of glowing mist. A shock of thunder eventually sweeps across the surface of the land beneath its path, rousing the birds from their trees, and shaking awake the startled hunters on the ground. The glowing descent is now accompanied by blasts of fire as its course is altered, and through the dazzling incandescence can be seen the vague shape of some kind of vessel. It slows, and directly beneath it a descending waft of hot air becomes a searing blast that incinerates

aquatics
seeker
hiver
parasite
host
fish-eater
antmen
spiketooth
hunter
slothmen
tree-dweller

Homo sapiens sapiens
Homo aquaticus
Homo caelestis
Homo sapiens machinadiumentum
Homo virgultis fabricatus
Homo glacis fabricatus
Homo silvis fabricatus
Homo campis fabricatus
Piscanthropus submarinus
Homo sapiens accessiomembrum
Homo mensproavodorum
Speluncanthropus
Moderator baiuli
Baiulus moderatorum
Homo dormitor
Homo vates
Alcearanthropus desertus
Homo nanus
Nananthropus parasitus
Penarius pinguis
Piscator longidigitus
Formifossor angustus
Acudens ferox
Harenanthropus longipis
Gigantanthropus arbrofagus
Arbranthropus lentus
Piscanthropus profundus

trees and undergrowth in a spreading circle. The vessel sinks into the boil of smoke and flame that is produced, and very gently it touches the ground.

The Earth's long period of innocence is over.

BUILDERS

Decades later the moving stars look down upon an altered countryside. At night the landscape glows. Buildings, the likes of which have not been seen on Earth for 5 million years, have appeared everywhere, but they are not like the structures of old. They are more like compact domes, or sealed cylinders and boxes, under pressure so that nothing of the natural planet can get in, or their contents get out.

What has happened? For 5 million years, since the extinction of technological man, the Earth has remained a reasonably natural place. Granted it has changed, with the enlarging and shrinking of icecaps, the rising and falling of sea levels, the reversal of the magnetic field, the creation of new mountain and volcanic island chains, the miniscule movements of continents, the replacement of forest by grassland, or grassland by desert, or desert by forest – but these changes have taken place with the incomprehensible slowness of the creep of natural processes. The only unnatural large-scale change has been the covering of the lowlands with algal mats for a time, but this also took many thousands of years to establish, and many thousands of years to disappear once more.

This is something different! In less than 100 years, since the coming of the lights in the sky, the world has been altered out of all recognition.

It would now be difficult to find a place on Earth that still had its natural vegetation. Everywhere is covered by the buildings. The first were brought down from above, but later they were made from the materials of the Earth itself. The carbon of the buried and fossilized algal mats was ripped up and used as fuel for all this change; and the minerals were gouged from the mountains and rendered down to make the great structures.

Even the oceans are partially covered. Huge rafts of floating cities, totally enclosed against the water and the air alike, mass on the surface of the waters.

Yet the builders, the occupants, the new masters, are never seen. They are so intolerant of conditions on the Earth's surface that they cannot expose themselves to it, so they remain in their sealed boxes. When they do venture forth, it is only in pressurized suits that disguise their true forms, confusing even the basic shapes of their suits by instruments and appendages that enable them to carry out some function.

It is difficult to realize that these beings have merely come home. Their ancestors left 5 million years ago in the great starships to seek out and colonize new worlds in other star systems. In order to build their ships, they drew upon the newly-developed science of genetic engineering, creating beings that could live and work in space, for the construction, and beings that could live and work in the oceans, for their provisioning. They took the knowledge of genetic engineering with them on their journey.

Far away from the Earth they found new planets, ones that could support life. However, the life that these planets could support was not the life of Earth, so from the offspring of the space-travellers new beings were engineered, ones that were able to breathe unfamiliar gas mixtures, subsist on the novel products of the new planetary surfaces, withstand the new forces of gravity and function under widely different atmospheric pressures than those to which their parents were accustomed.

Eventually these new creatures created their own civilizations on the far planets; but so great were the distances between them that they could never communicate with one another, and the development of each of the colonies proceeded independently. Most found conditions too harsh and perished. Some, however, established themselves and flourished.

Eventually a vastly different civilization, consisting of totally different creatures, and therefore based on unimaginable social and moral considerations, resumed exploration of the galaxy. On their travels, they met up with the other successful colonies, so completely changed that neither recognized the other as a cousin.

Now they have returned to Earth. Whether or not they recognize it as the planet from which their ancestors came is doubtful. If their bodies have changed totally, then so much more have their minds. It would be impossible for a mere Earth-bound imagination to understand the motives for their exploration, their attitudes to the life-forms they have found, or their long-term intentions.

However, the results of their exploration are easy to see. The planet's land surface has been changed beyond repair. Everywhere are the pressure-shelled settlements, the natural vegetation has been stripped away, to be used as raw materials for some process, and new gas mixtures are being generated and emitted from the factories to produce

5 MILLION YEARS HENCE

ENGINEERED PACK-ANIMAL

The second phase of biological engineering is exploitation. When applied to a planet this is known as 'terraforming'. Change and adaptation become secondary to whatever purpose the genetic engineers find important. Earth has not been exploited for 5 million years. When resources are abundant, methods of collecting and refining need not be sophisticated. The function of most of these altered creatures is as simple beasts of burden, able to operate within environments intolerable to their masters.

The atmosphere is being changed. With oxygen no longer present in quantities sufficient to keep *Homo sapiens*-based species alive, air-tanks and purification systems are essential. Control is by telepathic input direct to the central nervous system.

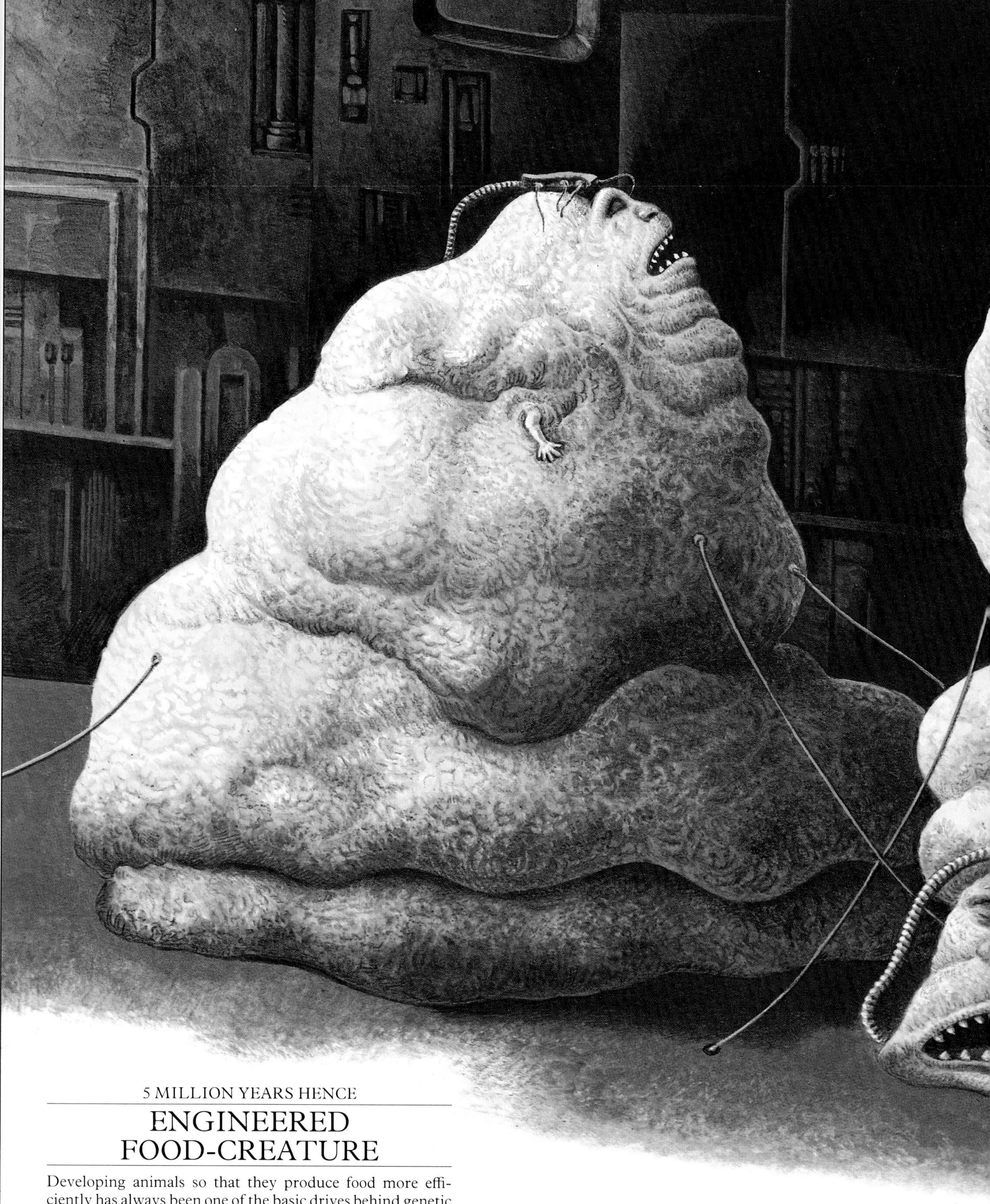

5 MILLION YEARS HENCE

ENGINEERED FOOD-CREATURE

Developing animals so that they produce food more efficiently has always been one of the basic drives behind genetic engineering. A food species may look grotesque – but then the natural forces of evolution often drive in a different direction to the consuming forces of science and civilization.

Penarius pinguis, *the parasite host, has been reduced to a mound of fat and flesh, fed by chemical nutrients. Harvesting devices remove meat as it is grown.*

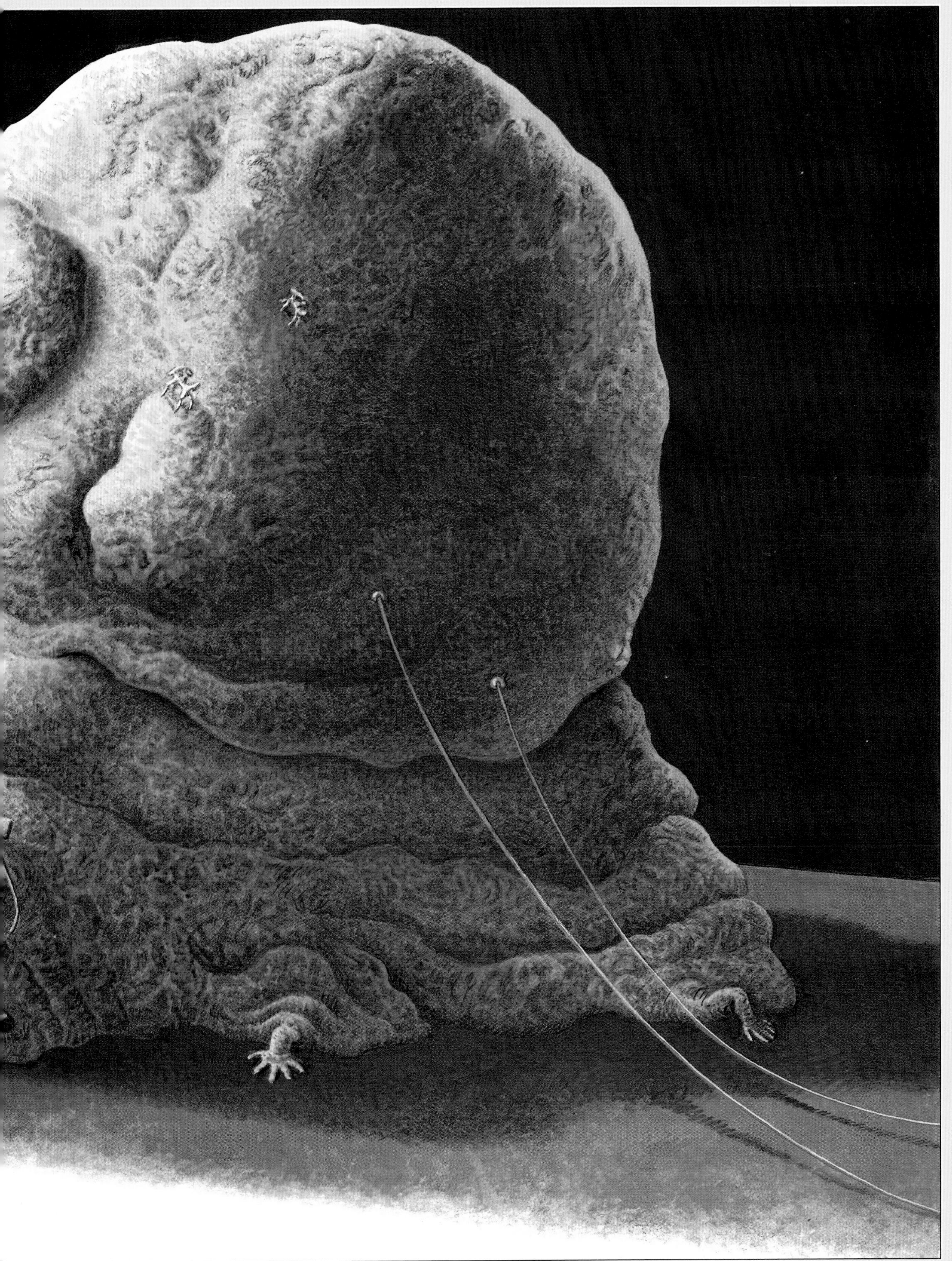

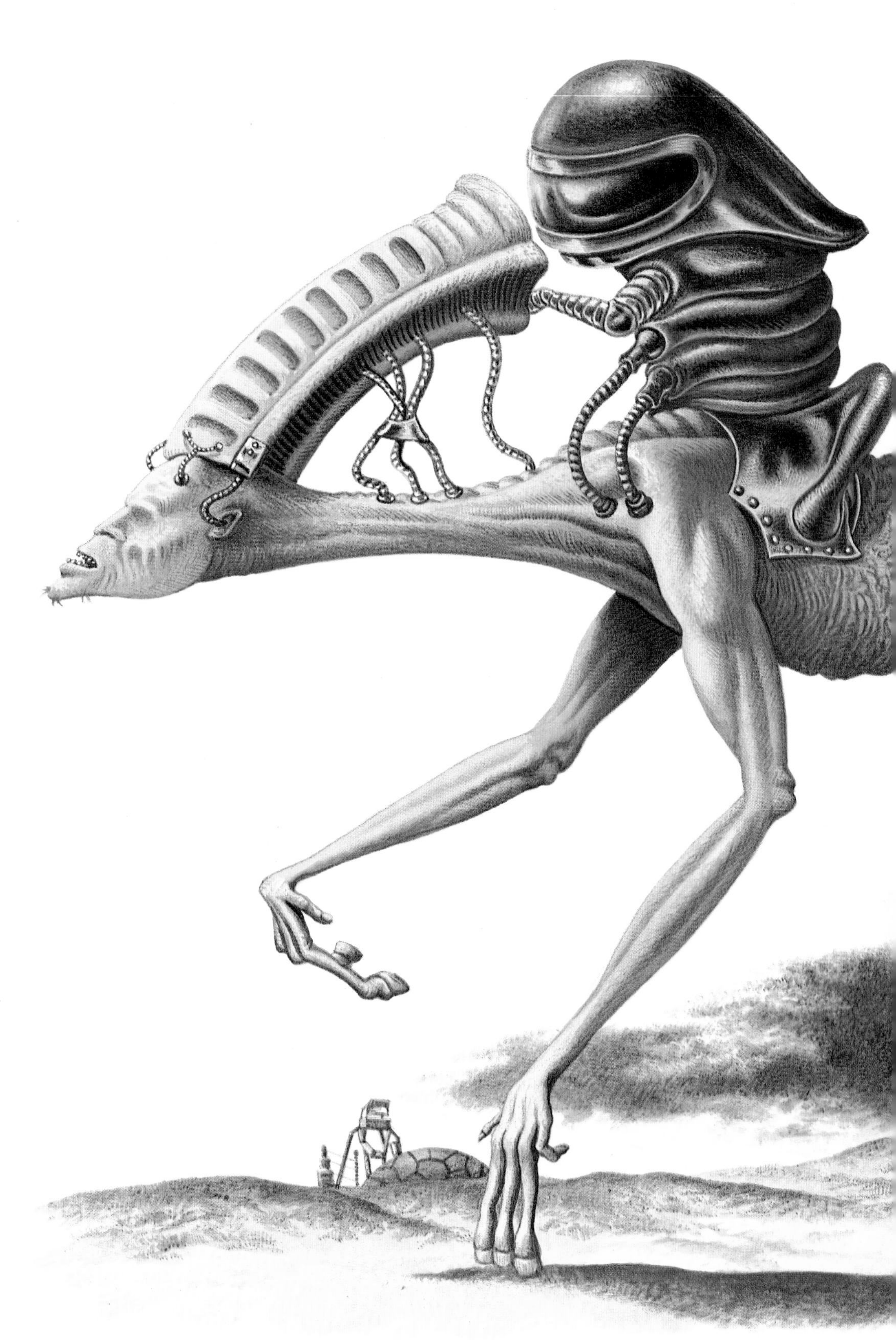

5 MILLION YEARS HENCE

JIMEZ SMOOT'S DESCENDANT

Descendant of *Homo sapiens sapiens* and the product of millions of years of genetic engineering and elective surgery, the newcomer is not yet at home in his new environment. The composition of the air can be changed but the unfamiliar atmospheric pressure presents greater problems. If the newcomer decides to stay, then further engineering will be essential. It was the constant need to withstand different gravities and breathe other atmospheres that led to one change being put on top of another; until genetically, psychologically and intellectually, the newcomer bears no resemblance to his ancestor, *Homo sapiens sapiens*.

Encased in a pressurized suit, Man's descendant sits astride a creature engineered from Homo virgultis fabricatus, *the temperate woodland-dweller. Direct telepathic control is exercised over the central nervous system of its mount.*

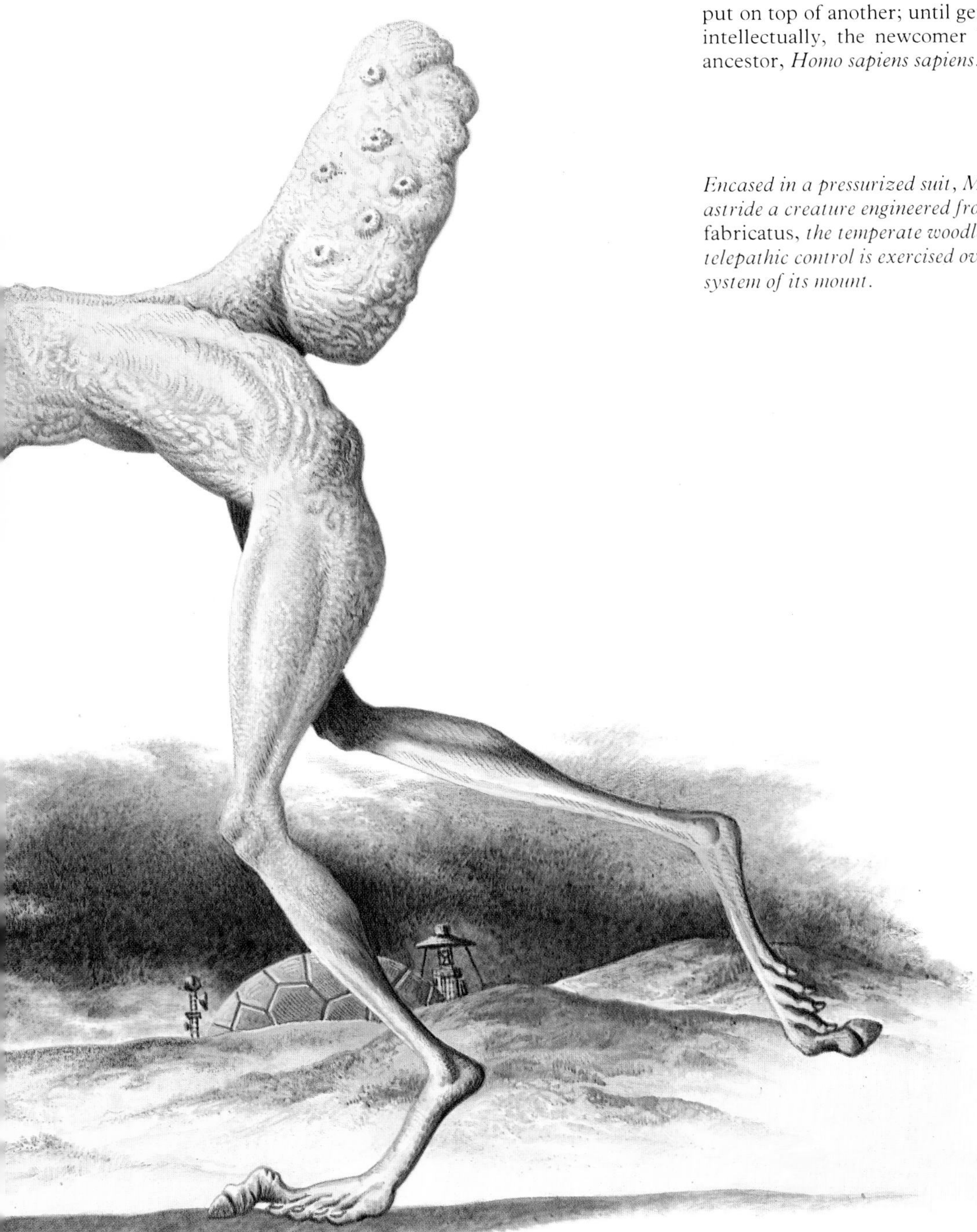

aquatics

seeker
hiver
parasite
host
fish-eater
antmen
spiketooth
hunter
slothmen
tree-dweller

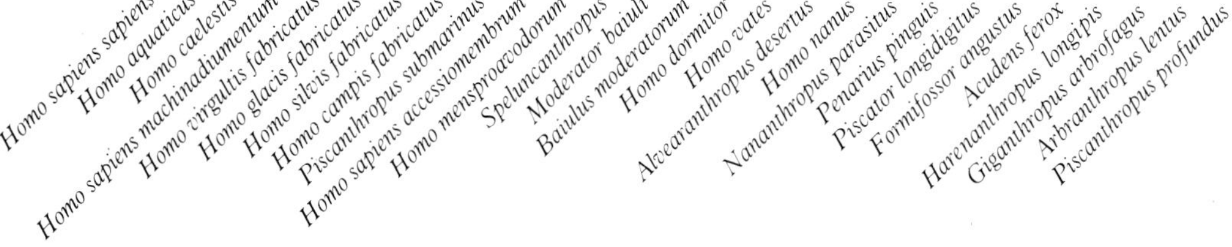

Even smaller forms are developed from Homo virgultis fabricatus, *to work intricate machinery in confined spaces. They are closest* Homo sapiens *comes to being a computer-aided soft machine.*

an altered atmosphere. A different people, using a different tongue, would call the process 'terraforming'.

The animal life has suffered terribly at their hands as well. With the spread of the new buildings and the destruction of the natural forests and vegetation, and the alteration of the atmosphere, most animals have perished. A few of the larger ones have been taken and used. The science of genetic engineering has been brought into play once more and any likely-looking animal has been altered to suit the newcomers' purposes.

Food is the main consideration. Many of the larger animals have been seen as excellent sources of protein for the newcomers. Some of the larger buildings now contain rows and rows of them, genetically refined, bloated, misshapen and unrecognizable. Huge mounds of fat and flesh grow in sturdy racks, fed by chemical nutrients pulsing through pumps and tubes connected directly into the tissues. Harvesting devices scour through the flesh, removing the meat and fat as it is grown. Only the presence of a few identifiable organs – shrivelled pulsing limbs and blind, gaping faces – show that these food-generators have been transformed from something that was once more noble.

Elsewhere other members of the local fauna have been engineered as work machines. These mostly retain their original shapes – two arms with prehensile hands, two legs with strong feet – but their heads are encased in metal and plastic boxes. Radio or telepathic receivers analyse their masters' wishes and stimulate the appropriate centres of the workcreatures' brains. Containers of their natural air keep them alive in the changing atmosphere. They have been developed in many sizes. Giants, larger than the extinct slothmen, carry heavy loads and put together the prefabricated parts of the buildings. Midgets, smaller than the parasite part of the extinct parasitehosts, manipulate the fine structures and operate in confined spaces. All sizes in between do the various other jobs. All work outside the domes and sealed cylinders is done by these beings.

When one of the newcomers ventures out, it does so encased in an anonymous armoured suit, and usually astride an engineered creature. The vaguely human limbs of this creature are now long and spindly, but still able to take its own weight and that of the being seated upon it. As with the workcreatures the top of its head is encased in a mechanical device that controls its brain directly.

These are all that remain of the creatures that once roamed the landscape of this planet.

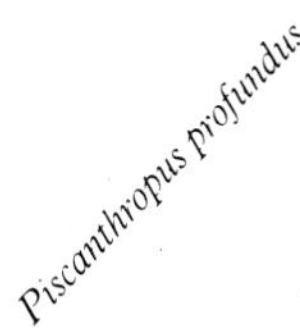

Homo sapiens sapiens
Homo aquaticus
Homo caelestis
Homo sapiens machinadiumentum
Homo cirgultis fabricatus
Homo glacis fabricatus
Homo silvis fabricatus
Homo campis fabricatus
Piscanthropus submarinus
Homo sapiens accessiomembrum
Homo mensproavodorum
Speluncanthropus
Moderator baiuli
Baiulus moderatorum
Homo dormitor
Homo vates
Alcearanthropus desertus
Homo nanus
Nananthropus parasitus
Penarius pinguis
Piscator longidigitus
Formifossor angustus
Acudens ferox
Harenanthropus longipis
Gigantthropus arbrofagus
Arbranthropus lentus
Piscanthropus profundus

EMPTINESS

Centuries later, all has changed again. The moving stars in the sky have gone. The newcomers have gone.

All living things on the surface of the Earth have also gone, leaving nothing but a harsh barren landscape littered with decaying and collapsing buildings. The sky is an unfamiliar colour, with the sunlight filtering through a foreign mix of gases and drifting pollutants. The rocks and exposed mountains are crumbling to sand and dust, as the new atmosphere – completely alien and incompatible with the physics of the Earth and the chemistry of its surface – slowly tries to find some kind of stability now that it has been deprived of the artificial systems and technologies that generated it and sustained it for such a short period of time.

The newcomers came, took what they wanted and departed once more, presumably to other planets that they could use for their purposes. What they left behind in the way of artefacts will slowly crumble away and disappear. The havoc that they wreaked in the atmosphere and surface of the land will take considerably longer to resolve itself.

IN THE END IS THE BEGINNING...

Throughout the oceans of the world there is a network of volcanic ridges. New crust is constantly being generated here, built up from hot material welling up from below the Earth's surface. The newly-formed crust is continually moving away from the ridge axes as even newer molten rock constantly forces its way up. This is the mechanism that moves the continents and alters the geography of the Earth.

The seawater along these ridges seeps into the newly-formed crust, is heated by the volcanic activity, dissolves many of the minerals that it finds there, and erupts once more as hot springs on the ridges. The instant cooling brings the minerals out of solution forming dense smoke-like clouds of suspended chemicals in the water around the vents.

The chemical energy present around these 'smokers' is immense, and bacteria thrive on it. Traditionally, organisms have relied on the energy of the sun to produce their food energy. Plants have used their chlorophyll to harness the sun's energy and use it to make their own food from the gases of the air and the minerals of the land. Animals have also eaten the food produced by the plants, and other animals have eaten the plant-eaters. All animals die and decay into the gases of the air and the minerals of the land, which are turned back into food by the plants. The sun's energy turns this wheel.

Around the smokers, well away from any light of the sun, the chemical energy is used by the bacteria to produce their own food. Simple, single-celled animals feed on the bacteria. More complex multi-celled animals feed on these, and so on. Gigantic worms and blind crabs thrive in these hot oases in the cold dark depths of the barren ocean – descended from remote ancestors that were once part of the sun-driven ecosystem far, far above them.

There is also another creature, only come to the smoky oases in the last few million years, and unnoticed by anything on the surface. This creature has a fish-like body that allows it to swim, and prehensile hands that help it to feel its way about and find its food. Its total blindness is no disadvantage in the solid blackness, but it has a sensitive organ on the top of the head, developed from an organ that its remote ancestors had forgotten, that is sensitive to heat and can give a temperature-based picture of the surroundings. More importantly, this creature has a brain that is complex enough to give it an intellect. This intellect tells it that something strange has taken place far above it.

Maybe someday it will be possible for its descendants to travel upwards, and even possible for them to live in conditions that are totally alien to it, if they can change enough.

Maybe someday....

FURTHER READING

NON-FICTION

Calder, N. *The Weather Machine* BBC Publications, London, 1966
Gregory, W.K. *Our Face, From Fish to Man* Capricorn Books, New York, 1965
Haldane, J.B.S. *Possible Worlds* Evergreen Books, London, 1940
Lunan, D. *Man and the Planets* Ashgrove Press, Bath, 1983
Nicholls, P. (ed) *The Science in Science Fiction* Roxby Press, London, 1983
Pain, S. 'No Escape from the Global Greenhouse' *New Scientist*, vol 120, no. 1638, 12 November 1988
Ridpath, I. *Life Off Earth* Granada, London, 1983
Stapleford, B. *Future Man* Roxby Press, London, 1984

FICTION

The works of fiction that depict the future of Man and the changes that he may go through are legion. Here are just a few.

Adams, D. *The Hitch-Hiker's Guide to the Galaxy* Pan, London, 1979
Aldiss, B. *Hothouse*, Faber & Faber, London, 1962
Aldiss, B. *Canopy of Time* Faber & Faber, London, 1959
Bass, J.T. *Godwhale* Eyre Methuen, London, 1974
Brunner, J. *The Sheep Look Up* Dent, London, 1974
Budrys, A. *Who* Penguin, London, 1958
Harrison, H. *Make Room* ! *Make Room*! Doubleday, New York, 1966
Huxley, A. *Brave New World* Chatto & Windus, London, 1932
Pohl, F. *Man Plus* Gollancz, London, 1976
Simak, C. *City* Weidenfeld & Nicholson, London, 1954
Vance, J. *The Dragon Masters* Galaxy, 1963
Vinge, V. *Marooned in Real Time* Simon & Schuster, New York, 1986
Wells, H.G. *The Time Machine* Heinemann, London, 1895
Wells, H.G. *The Island of Dr Moreau* Heinemann, London, 1896

CAST I

Homo sapiens sapiens	Jimez Smoot Kyshu Kristaan Seralia Kristaan Fiffe Floria Hamstrom Harla
Homo sapiens machinadiumentum	Haron Solto Greerath Hulm Hueh Chuum Bearnida Kule Taaran Relia Hoolann Carahudru
Homo sapiens addomembrus	Klimasen Yamo Durian Skeel
Homo aquaticus	Piccarblick
Homo caelestis	Cralym
Homo virgultis fabricatus	Hoot Rumm Coom Snatch Trancer
Homo glacis fabricatus	Knut
Homo silvis fabricatus	Pann
Homo campis fabricatus	Gram Larn
Piscanthropus submarinus	Ghloob
Baiulus moderatorum	Oyo
Homo mensproavodorum	Hrusha Vass Kroff

CAST II

Bearnida	*Homo sapiens machinadiumentum*
Chuum, Hueh	*Homo sapiens machinadiumentum*
Carahudru	*Homo sapiens machinadiumentum*
Coom	*Homo virgultis fabricatus*
Cralym	*Homo caelistis*
Floria, Fiffe	*Homo sapiens sapiens*
Floria, Hamstrom	*Homo sapiens sapiens*
Floria, Harla	*Homo sapiens sapiens*
Ghloob	*Piscanthropus submarinus*
Hoolann, Relia	*Homo sapiens machinadiumentum*
Hrusha	*Homo mensproavodorum*
Hulm, Greerath	*Homo sapiens machinadiumentum*
Klimasen	*Homo sapiens addomembrus*
Knut	*Homo glacis fabricatus*
Kristaan, Kyshu	*Homo sapiens sapiens*
Kristaan, Seralia	*Homo sapiens sapiens*
Kroff	*Homo mensproavodorum*
Larn	*Homo campis fabricatus*
Oyo	*Baiulus moderatorum*
Pann	*Homo silvis fabricatus*
Piccarblick	*Homo aquaticus*
Rumm	*Homo virgultis fabricatus*
Skeel, Durian	*Homo sapiens addomembrus*
Smoot, Jimez	*Homo sapiens sapiens*
Snatch	*Homo virgultis fabricatus*
Taaran, Kule	*Homo sapiens machinadiumentum*
Trancer	*Homo virgultis fabricatus*
Vass	*Homo mensproavodorum*
Yamo	*Homo sapiens addomembrus*

INDEX

Figures in *italics* refer to illustrations.